Back to Basics in Palaeontology

Back to Basics in Palaeontology

Guest Editor

Hua Stephane

MDPI

Basel • Beijing • Wuhan • Barcelona • Belgrade • Novi Sad • Cluj • Manchester

Guest Editor
Hua Stephane
Odyssée Villers Paléospace
Villers-sur-Mer
France

Editorial Office
MDPI AG
Grosspeteranlage 5
4052 Basel, Switzerland

This is a reprint of the Special Issue, published open access by the journal *Life* (ISSN 2075-1729), freely accessible at: https://www.mdpi.com/journal/life/special_issues/GT8H3M2XII.

For citation purposes, cite each article independently as indicated on the article page online and as indicated below:

Lastname, A.A.; Lastname, B.B. Article Title. *Journal Name* **Year**, *Volume Number*, Page Range.

ISBN 978-3-7258-8003-4 (Hbk)
ISBN 978-3-7258-8004-1 (PDF)
https://doi.org/10.3390/books978-3-7258-8004-1

Cover image courtesy of Hua Stephane

Contents

About the Editor

Hua Stephane

Hua Stephane holds a PhD in paleontology and is a computer scientist by trade; Doctor of the University of Paris 6 and free researcher at Paléospace, paleontologist specializing in saltwater crocodiles. As a child, since the age of 6, I wanted to do paleontology. I knew I was destined for this. I continued my studies towards this ultimate goal: paleontology. By chance during my studies, I had the chance to meet Éric Buffetaut, a renowned paleontologist and director of research at the CNRS. One day he said to me, "There are many researchers on dinosaurs but no one on crocodiles! So if you're interested..." And that's when I said to myself, "Banco!" I wanted a big predator with big teeth! So crocodiles, I thought, were cool. Over time, these animals may have been insectivores, carnivores, marine, terrestrial, etc. They even survived the Cretaceous-Tertiary crisis! It's a fascinating group. The more you study it, the more you want to discover more. Paleontology is, in a way, abstracting oneself from time. In my opinion, a paleontologist is defined above all by his curiosity and his desire to understand. You are always surprised; there are new techniques or discoveries that appear continuously! Beyond reconstructing part of the past, the beautiful thing about paleontology is that when you write articles, they have to transcribe a logic in its pure state and require a certain humility. I think a "good" paleontologist is a researcher who has written a paper and decades later is still right. To sum up, what I like about this job is the somewhat eternal and logical side.

Preface

In this reprint, we have compiled paleontological papers with a dedicated interest in its history, its interaction with the other coexisting taxas and its paleoecosystem.

How do you define your taxa? How could it be differentiated from the rest of its family and eventually its originality? What makes some fossils exceptional? These papers have tried to answer all these questions. We are happy to publish them and we hope you are happy to discover them.

Hua Stephane
Guest Editor

life

Article

Calamitean Cones and Their In Situ Spores from the Pennsylvanian Limnic Basins of the Czech Republic

Jiří Bek * and Jana Votočková Frojdová

Institute of Geology, Academy of Sciences of the Czech Republic, Rozvojová 236, 16500 Prague, Czech Republic; frojdova@gli.cas.cz
* Correspondence: bek@gli.cas.cz

Abstract: This paper describes the in situ spores of the *Calamospora* type, macerated from sixty-one specimens of calamitean cones belonging to sixteen species of genera, such as the *Palaeostachya*, *Macrostachya*, *Calamostachys*, and *Huttonia* from the Pennsylvanian Czech Republic period, specifically from the Moscovian/Kasimovian ages (i.e., Duckmantian-Stephanian). The in situ spores were compared to twenty dispersed species of *Calamospora*. The majority of spores were microspores; however, some cones yielded both micro- and megaspores. Morphological variations of the in situ spores, including the diameter, labrum, contact area, ontogenetic stages, and secondary folds of the exine, are described, including their importance for the classification of calamospores. The relationships of *Elaterites*, *Pteroretis*, *Vestispora*, and some monopseudosaccate spores are discussed. All Paleozoic *Calamospora*-producing parent plants are summarized.

Keywords: sphenophytes; calamites; in situ spores; *Calamospora*; Pennsylvanian

1. Introduction

Sphenophytes are an important group of both extant and extinct plants. The first specimens come from the Devonian period [1–3] but were never abundant. The "Golden age" for sphenophytes was the Carboniferous period, especially in Pennsylvanian times when arborescent forms reached up to 30 m [4]. After the Permian period, their diversity gradually declined until recent times. We know of only one recent genus, *Equisetum* Linnaeus, which has fifteen to eighteen species [4]. Some phylogenetic analyses [5] suggest that sphenophytes and ferns can be a monophyletic group [6]. Paleozoic sphenophytes are divided into two main groups: calamiteans and sphenophyllaleans. All Paleozoic calamiteans produced spores of the *Calamospora* type. Sphenophyllaleans can be palynologically divided into 6–7 groups, and only one of them produced *Calamospora* microspores [7].

Calamospora Schopf et al. is a spore genus consisting of mio- and megaspore species with long stratigraphical ranges from the Silurian [8] to Tertiary periods [9]. It is the morphologically simplest spore type, with circular to oval amb, simple trilete rays, and laevigate exine. Many morphologically simple spore types were produced by parent plants of different affinities, including *Calamospora* [7]. This is the reason why *Calamospora* has no stratigraphical and palaeoecological significance.

The genus was established [8] with the species type *C. hartungiana* Schopf et al. The number of classification criteria is low [10,11]. These include the contact area, length of rays of the trilete mark, labrum, number, size, shape, position of the secondary folds of the exine, and diameter. Biologically, calamospores were both male and female spores that cannot be distinguished morphologically but only by their diameter; 200 μm is an artificial "boundary" for the recognition of calamospores into male and female spores [7]. However, this boundary size does not always correspond with the biological function of calamospores. The main producers of *Calamospora* calamitean cones were the genera *Calamostachys* Schimper, *Palaeostachya* Weiss, *Macrostachya* Schimper, and *Paracalamostachys*

Weiss. Other parent plants mainly include noeggerathialeans and certain sphenophyllaleans [7]. The relationship of *Calamospora* from *Elaterites* Wilson, *Vestispora* (Wilson & Hoffmeister) Wilson & Venkatachala, and some monopseudosaccate forms of the *Remysporites/Auroraspora/Perotrilites* type has been discussed [10,12]. This paper describes sixty-one specimens of four calamitean *Calamospora*-producing genera from the Kladno-Rakovník, Radnice, and the Czech part of the Intra-Sudetic basins of the Czech Republic. All palynological aspects of *Calamospora* are summarized, including the morphology, ontogeny, and origin. Their classification is taken from [13]. The most important papers on Carboniferous calamites were published by [13,14], which documented in situ spores [15].

2. Geological Setting

Specimens of calamitean cones come from the Kladno-Rakovník Basin, Moscovian-Kasimovian (Upper Duckmantian-Lower Stephanian), Moscovian (Upper Duckmantian) of the Radnice Basin (Figure 1) and Bashkirian-Moscovian of the Intra-Sudetic Basin (Langsettian-Duckmantian, Stephanian, Figure 1).

The Central and Western Permo-Carboniferous limnic basins of the Czech Republic consist of the Kladno-Rakovník, Pilsen, and Radnice basins and a part of the Žihle Basin. The stratigraphical range of these sediments is Moscovian to Kasimovian (Duckmantian to Stephanian) and consists of the Kladno, Týnec, Slaný, and Líně formations. Sediments are up to 1400 m thick [16].

The oldest sediments of the Kladno-Rakovník Basin are early Moscovian (Duckmantian), and the youngest are Gzhelian (upper Stephanian) in age (Table 1).

Table 1. Stratigraphical position of studied calamitean cones.

Period	Age	
303 Ma	Kasimovian	Stephanian
Pennsylvanian 307 Ma	Moscovian	Asturian
		Bolsovian
315 Ma	Bashkirian	Duckmantian
		Langsettian

The Radnice Basin is a small erosional remnant west of the Pilsen Basin (Figure 1). Sedimentation started in the early Moscovian (late Duckmantian) and lasted until the early Permian (Asselian). Stratigraphically, the oldest sediments of the Radnice Member consist of the Radnice group of coals with lower and upper Radnice coal seams and the fossiliferous Whetstone Horizon between them. The Whetstone Horizon is well-known for its rich plant fossils, including the specimens described herein. The lower part of the overlying Nýřany Member is preserved following post-Permian erosion [16,17]. The thickness of members is variable due to the fluvial valleys incised by rivers into the basement.

The Intra-Sudetic Basin belongs to a large complex of Late Palaeozoic continental basins of the Bohemian Massif (Figure 1). The majority of the area (1800 km^2) is in Poland, with only about a third in the Czech Republic (Figure 1). The sedimentation started in Serpukhovian (late Viséan) and, including several hiatuses, continued for over 80 Ma until the Triassic [18,19]. A detailed overview of the depositional history is described by [17,19]. Serpukhovian (Late Viséan to early Namurian) alluvial deposits of the Blazkow Formation are the oldest in the Czech part of the basin [19]. After a short break, sedimentation resumed in the Bashkirian (late Namurian-middle Westphalian) times by fluvial facies of the Žacléř Formation. After another short break [18], red beds of the Moscovian (late Westphalian to middle Stephanian) Odolov Formation were deposited. The overlying Chvaleč Formation, with other red bed complexes, spans the late Gzhelian (Stephanian) and early Permian (Asselian) periods. The profile continues with the early Permian fluvial red bed [18].

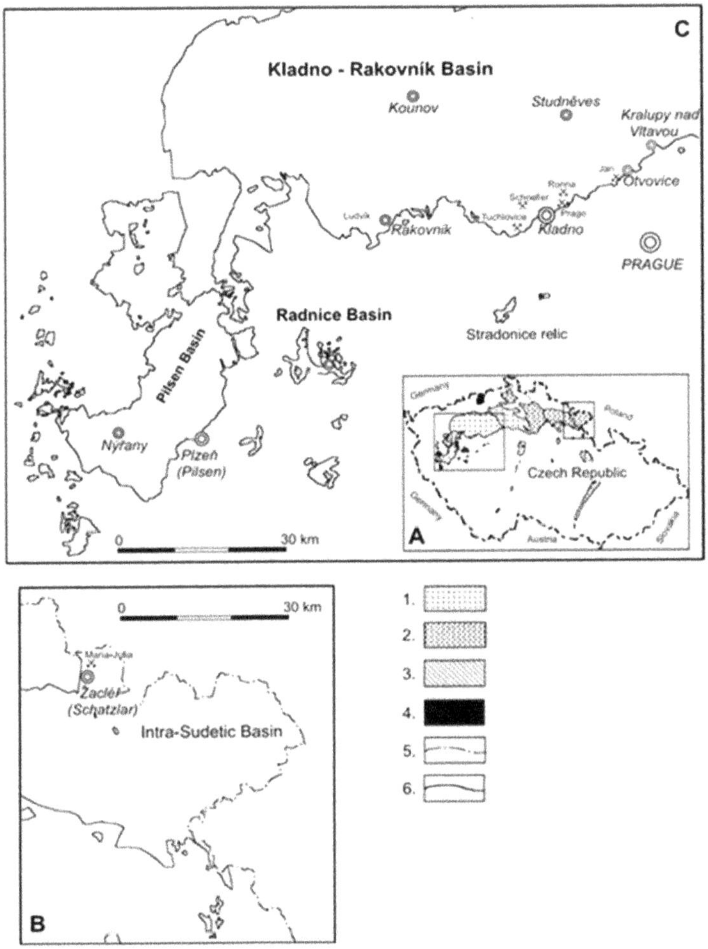

Figure 1. (**A**) Late Palaeozoic continental basins of the Czech Republic. (**B**) Position of localities in the Intra-Sudetic Basin. (**C**) Localities in the basins in central and western Bohemia. Explanation: 1. Central and Western Bohemian Upper Palaeozoic Basins; 2. Lusatian Upper Palaeozoic Basins; 3. Grabens; 4. Krušné Hory Upper Palaeozoic; 5. state border; and 6. present-day limit of continental basins.

3. Material and Methods

The specimens labeled E are housed in the National Museum, Prague, Czech Republic. Some specimens are from the collection of S. Opluštil, Faculty of Sciences, Prague, Czech Republic. A Nikon Eclipse BX51 light microscope was used for the study of palynological slides. Digital photomicrographs of all the in situ microspores are stored in the Laboratory of Palaeobiology and Palaeoecology, Institute of Geology v.v.i., Academy of Sciences, Prague. Spores were recovered by dissolving small portions of sporangia with the aid of nitric acid for 24–40 h and in KOH for 1 h. The majority of spores were mounted in glycerine jelly for a direct microscopic examination. A minority of them were coated with gold and prepared for SEM observation. The terms used for the descriptions of the in situ spores were obtained from the latest edition of the Glossary of Pollen and Spore Terminology [20]. In situ, the spores were compared according to the system of classification of dispersed spores [21–23].

4. Systematic Paleontology

Equisetales
Calamitaceae
Palaeostachya Weiss 1876
Type species: *Palaeostachys elongata* (Presl) Weiss 1876
Palaeostachya ettingshausenii Kidston 1903
Figures 2–4

Figure 2. *Palaeostachya ettingshausenii* Kidston. (**a**–**h**) Specimen No. E3622. Maria Julia Mine near Třemošín, Intra-Sudetic Basin, Stephanian. (**a**) General view. Scale bar 20 mm. (**b**,**c**) In situ megaspores of the *Calamospora* type. All ×100. (**d**,**e**) In situ microspores of the *Calamospora* type. Note the number of irregular folds of exine. All ×500. (**f**) In situ microspore of the *Calamospora* type. Note that the rays of the trilete mark reach minimally three-quarters of the radius. ×500. (**g**) In situ microspore of the *Calamospora* type. Note that the rays of the trilete mark reach about three-quarters of the radius. ×500. (**h**) Immature microspore. ×500. (**i**) In situ microspore of the *Calamospora* type. ×500. (**j**) General view of specimen No. E2414. Mayray Mine near Kladno, Kladno-Rakovník Basin, Moscovian. Scale bar 20 mm. (**k**,**l**) In situ microspores of the *Calamospora* type. ×500.

Figure 3. *Palaeostachya ettingshausenii* Kidston. (**a**) General view of the whole specimen No. E3624, Mayray Mine near Kladno, Kladno-Rakovník Basin, Moscovian. Natural size. (**b**) Megaspore of the *Calamopora* type. ×100. (**c–f**) Microspores of the *Calamospora* type. (**e**) Note one major fold of exine covers the spore body. (**f**) Note a secondary fold of exine parallels the margin of the spore. All ×500.

Material: The specimens are from the S. Opluštil's collection and are as follows: Nos. 935 and 1451 are from the Tuchlovice Mine, Kladno-Rakovník Basin, Czech Republic, Bashkirian (Upper Duckmantian); E2414, E3624, E3626, E2412, and E3623 are from the Mayray Mine, Kladno; E3618 is from the Max Mine, Kladno; E3635 is from the Ronna Mine, Kladno; and E3596 is from the Břasy locality. All are from the Kladno-Rakovník Basin. The following specimens are from the Intra-Sudetic Basin as follows: E3622 is from the Maria-Julia Mine near Třemošín, Kasimovian (Stephanian); and E2500 is from the Žacléř locality, with an unknown stratigraphic position.

A description of the in situ spores is as follows: The trilete spores are circular, subtriangular, or oval in amb. Two kinds of spores probably represent the microspores (Figures 2d–i,k,l, 3c–f, and 4c,d,f–i) and megaspores (Figures 2b,c, 3b, and 4a,b). The laevigate exine of the microspores is up to 4 μm thick, and that of the megaspores is 6–9 μm thick. The rays of the trilete mark reach a third to two-thirds of the radius. Secondary folds of the exine possess variable sizes, shapes, positions, and numbers. Three size categories of the in situ spores were recognized (Table 2): the first with a size range of 44 (71) 171 μm (the majority of specimens), and the second is 80 (139) 181 μm; the megaspores are 234 (448) 708 μm. The contact area is sometimes developed. The in situ microspores can be compared to the dispersed miospore species *Calamospora microrugosa* (Ibrahim) Schopf et al., *C.* cf. *microrugosa*, *C. pedata* Kosanke, *C.* cf. *pedata*, *C. pallida* (Loose) Schopf et al., *C.* cf. *pallida*, *C.* cf. *liquida*, *C. breviradiata* Kosanke, and *C. hartungiana* Schopf et al., and the megaspores to the species *C. perrugosa* (Loose) Schopf et al.

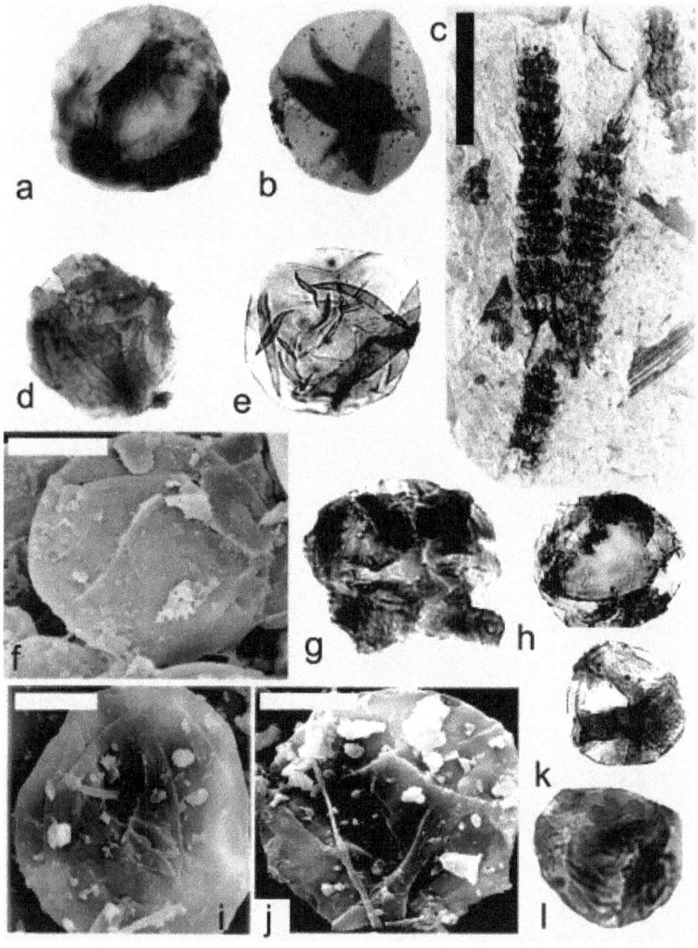

Figure 4. *Palaeostachya ettingshausenii* Kidston. (**a**,**b**) In situ megaspores of the *Calamospora* type isolated from specimen No. 3624, Mayray Mine near Kladno, Kladno-Rakovník Basin, Moscovian. (**c**) Specimen No. E935, Tuchlovice Mine near Kladno, Kladno-Rakovník Basin. Scale bar 175 mm. (**d**,**g**,**h**,**k**,**l**) In situ microspores of the *Calamospora* type. Note that irregular fragments of probable tapetal tissues cover the microspore body. All ×500. (**e**) In situ microspore of the *Calamospora* type. ×500. (**f**,**i**,**j**) In situ microspores of the *Calamospora* type. SEM. Scale bars 20 μm.

Table 2. The size of the in situ spores of *Palaeostachya ettingshausenii* Kidston.

Specimen Number	Diameter of In Situ Spores (μm)
E 2500—basal part	55.0 (67.64) 84.0
apical part	44.0 (62.3) 72.0
E 2412 right cone, basal part	84.0 (121.1) 181.0
apical part	62.0 (105.77) 131.0
left cones	461.0 (576.75) 708.0
E 3623	56.0 (90.27) 121.0
E 3625	58.0 (92.33) 126.0
E 2414	48.0 (71.19) 101.0
E 1117 upper cone, basal part	49.0 (64.6) 74.0
middle part	48.0 (61.22) 79.0
lower cone	43.0 (63.2) 74.0
E 3596	56.0 (92.25) 131.0
E 3622	76.0 (84.52) 94.0
	234.0 (398.58) 602.0
E 3624	390.0 (494.38) 621.0
E 3618	330.0 (420.88) 592.0
	54.0 (69.99) 79.0
E 3626	75.0 (119.5) 151.0
935 left cone, basal part	60.0 (83.1) 100.0
apical part	52.0 (67.44) 74.0
right cone, basal part	71.0 (75.97) 80.0
apical part	58.0 (64.22) 70.0
1451 left cone, basal part	64.0 (71.63) 89.0
right cone, basal part	60.0 (67.81) 80.0
apical part	52.0 (59.32) 70.0

Palaeostachya distachya (Sternberg) Jongmans 1911

Figures 5 and 6a–e

Material: E3608, E3598, and E3604 are from Hnidousy, near Kladno; and E1132, E3599, E3603, E1132, and E1134 are from the Ronna Mine. All of them are from the Kladno-Rakovník Basin, Bashkirian-Moscovian (Upper Duckmantian/Lower Bolsovian).

A description of the in situ spores is as follows: The trilete spores are circular, subtriangular, or oval in amb. The laevigate exine of the microspores is up to 3 μm thick. The rays of the trilete mark reach a third to two-thirds of the radius. Secondary folds of the exine possess variable sizes, shapes, positions, and numbers. The size range is 34 (65) 119 μm (Table 3). The contact area (Figure 5b–d) and labrum (Figures 5f,j,l and 6d,e) are sometimes developed. The in situ microspores can be compared to the dispersed miospore species *Calamospora mutabilis* (Loose) Schopf et al., *C.* cf. *pedata*, *C. minuta* Bharadwaj, C. cf. *minuta*, *C. hartungiana*, *C.* cf. *hartungiana*, *C. breviradiata*, C. cf. *breviradiata*, and *C saariana* Bharadwaj.

Table 3. Size of the in situ spores of *Paleostachya distachya*.

Specimen Number	Diameter of In Situ Spores (μm)
E 3599	54.0 (64.36) 79.0
E 3603	55.0 (63.14) 76.0
E 3500	55.0 (66.2) 78.0
E 3608	36.0 (60.82) 108.0
E 3598	41.0 (62.6) 99.0
E 1132	33.0 (76.4) 92.0
E 1134	59.0 (90.58) 119.0
E 3604	71.0 (87.38) 111.0

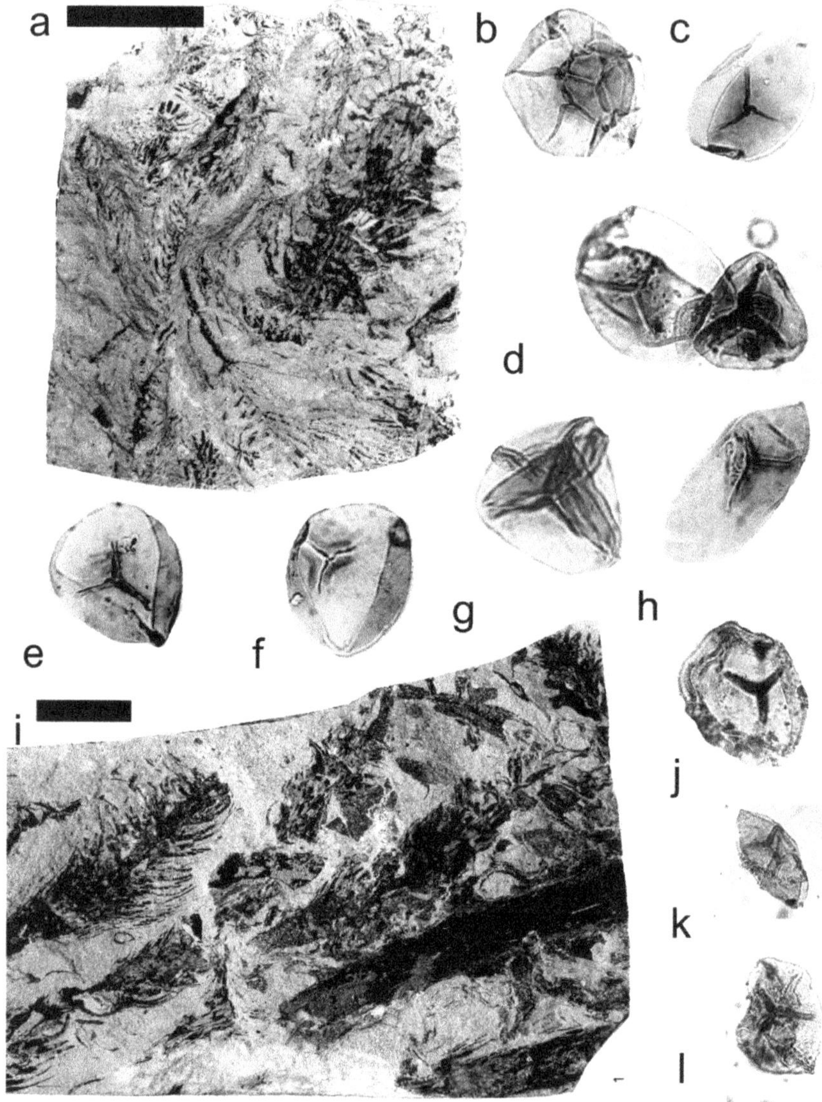

Figure 5. *Palaeostachya distachya* (Sternberg) Jongmans. (**a**) Specimen No. E1132, Ronna Mine near Kladno, Kladno-Rakovník Basin, Moscovian. Scale bar 20 mm. (**b**) In situ microspore of the *Calamospora* type. Note the dark contact area. ×500. (**c**) In situ microspore of the *Calamospora* type. Note that one major fold covers the microspore body. ×500. (**d**) Two in situ microspores of the *Calamospora* type. The left specimen represents a mature microspore, while the right specimen is relatively immature. ×500. (**e**) In situ microspore of the *Calamospora* type. Note the two major folds of exine. ×500. (**f**) In situ microspore of the *Calamospora* type. Note the dark, thickened labrum. ×500. (**g**) In situ microspore of the *Calamospora* type. ×500. (**h**) In situ microspore of the *Calamospora* type. Note the dark labrum. ×500. (**i**) Specimen No. E3608, Hnidousy near Kladno, Kladno-Rakovník Basin, Moscovian. Scale bar 20 mm. (**j**) In situ microspore of the *Calamospora* type. Note the dark, thin labrum. ×500. (**k**,**l**) In situ microspores of the *Calamospora* type. All ×500.

Palaeostachya pedunculata Williamson 1874
Figure 6f–k

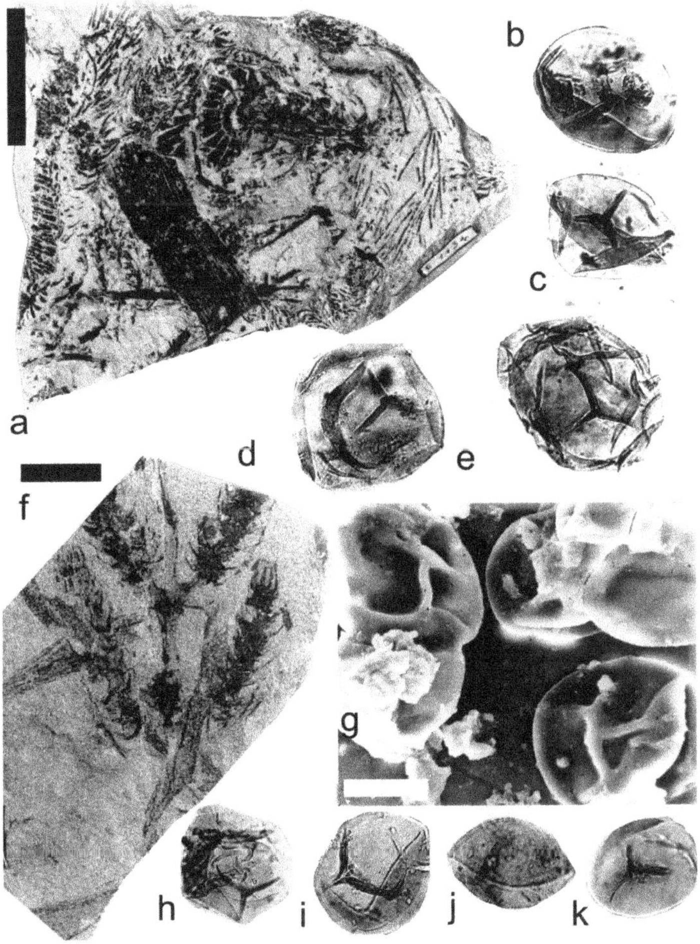

Figure 6. (**a–e**) *Palaeostachya distachya* (Sternberg) Jongmans. (**a**) Specimen No. E3608, Hnidousy near Kladno, Kladno-Rakovník Basin, Moscovian. Scale bar 20 mm. (**b**) In situ microspore of the *Calamospora* type. Note that one major fold covers the microspore body. ×500. (**c**) In situ microspore of the *Calamospora* type. Note the dark labrum. ×500. (**d**,**e**) In situ microspores of the *Calamospora* type. All ×500. (**f–k**) *Palaeostachya pedunculata* Williamson. (**f**) Specimen No. E1125, Ronna Mine Kladno, Kladno-Rakovník Basin, Moscovian. Scale bar 10 mm. (**g**) In situ microspores of the *Calamospora* type. Note the elevated labrum. SEM. Scale bar 25 µm. (**h–k**) In situ microspores of the *Calamospora* type. All ×500.

Material: No. E1125, from the Ronna Mine near Kladno, Kladno-Rakovník Basin, Bashkirian (Upper Duckmantian).

A description of the in situ spores is as follows: The trilete spores are circular, subtriangular, or oval in amb. The laevigate exine of the microspores is up to 2 µm thick. The rays of the trilete mark reach half of the radius. Secondary folds of the exine possess variable sizes, shapes, positions, and numbers (Figure 6g–k). The size range is 51 (70) 89 µm. The labrum is sometimes developed (Figure 6g,i). The in situ microspores can be compared to the dispersed miospore species *Calamospora* cf. *pedata*, *C.* cf. *microrugosa*, and *C.* cf. *straminea*.

Palaeostachya elongata (Presl) Weiss 1976
Figure 7a–o

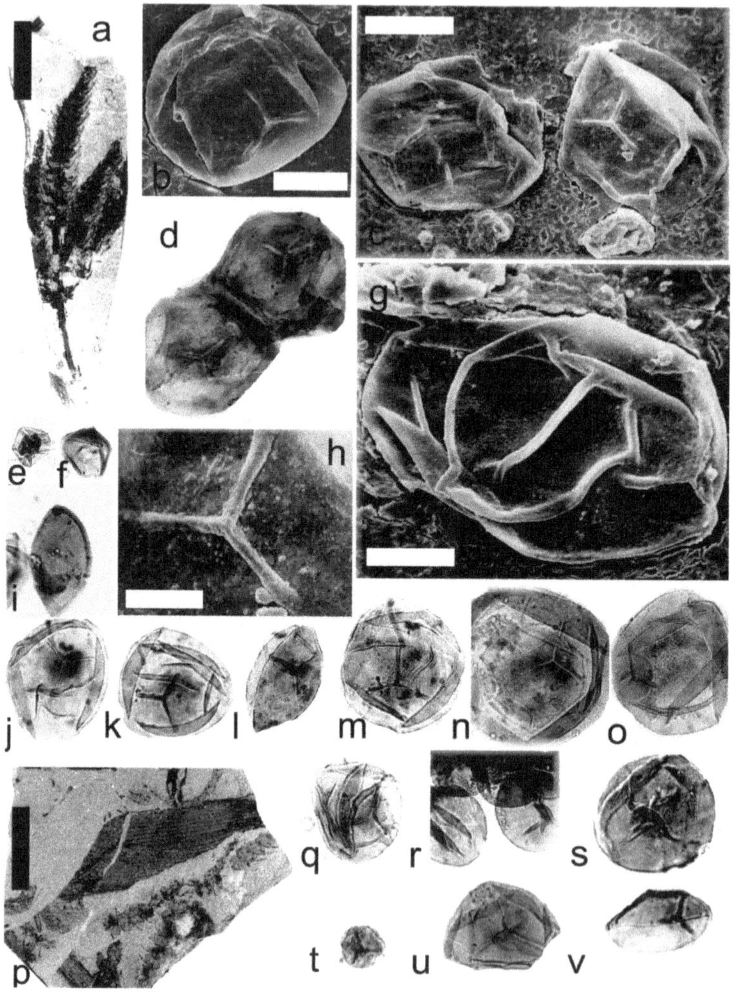

Figure 7. (**a–o**) *Palaeostachya elongata* (Presl) Weiss. (**a**) Specimen No. E1122, Kladno Locality, Kladno-Rakovník Basin, Moscovian. Scale bar 20 mm. (**b**,**c**) In situ microspores of the *Calamospora* type. SEM, scale bar 20 μm. (**d**) Two in situ microspores of the *Calamospora* type with fragments probably of tapetal tissues. ×500. (**e**,**f**) Two immature in situ microspores. All ×500. (**g**) In situ microspore of the *Calamospora* type. SEM, scale bar 20 μm. (**h**) Elevated labrum of in situ microspore of the *Calamospora* type. Detail of Figure (**c**) (right specimen). SEM, scale bar 10 μm. (**i**) In situ microspore of the *Calamospora* type. Note one major fold covering the body. ×500. (**j**,**k**,**m–o**) In situ microspores of the *Calamospora* type. Note the dark contact area. All ×500. (**l**) In situ microspore of the *Calamospora* type. Note one major fold covers half of the spore body. ×500. (**p–v**) *Palaeostachya gracillima* Weiss. (**p**) Specimen No. E1119. Kladno locality, Kladno-Rakovník Basin. Moscovian. Scale bar 20 mm. (**q–s**,**u**) In situ microspores of the *Calamospora* type. All ×500. (**t**) Immature in situ microspore. ×500. (**v**) In situ microspore of the *Calamospora* type. Note one major fold of the exine covers half of the spore body. ×500.

Material: No. E1122 is from Bashkirian (Upper Duckmantian), Kladno locality, Kladno-Rakovník Basin; No. E1121 is from Moscovian (Bolsovian), Rakovník locality, Kladno-Rakovník Basin; Nos. E3607 and E 3616 are from Moscovian (Bolsovian), the Šamotka locality, Věnec Coal Seam near Lubná; No. E3631 is from Bashkirian (Langsettian-Duckmantian) of the Intra-Sudetic Basin, Maria-Julia Mine near Žacléř.

A description of the in situ spores is as follows: The trilete spores are circular, subtriangular, or oval in amb. The laevigate exine of the microspores is up to 2 µm thick. The rays of the trilete mark reach a third to two-thirds of the radius. Secondary folds of the exine possess variable sizes, shapes, positions, and numbers (Figure 7b–d,j–o). The size range is 60 (112) 141 µm (Table 4). The contact area and labrum are up to 6 µm in size and sometimes developed (Figure 7c,h,j,l,n,o). Sometimes, fragments of the perispore can occur on the surface of the exine (Figure 7d). The in situ microspores can be compared to the dispersed miospore species *Calamospora hartungiana*, *C.* cf. *hartungiana*, *C. falkenbergensis* Venkatachala & Bharadwaj, and *C. mutabilis*. It is possible that some spores that are classified as *Calamospora falkenbergensis*, *C.* cf. *hartungiana*, and *C. mutabilis*, with size ranges of 82 to 152 µm, may be megaspores.

Table 4. Size of the in situ spores of *Palaeostachya elongata*.

Specimen Number	Diameter of In Situ Spores (µm)
E 3607	84.0 (112.2) 123.0
E 3631	35.0 (42.35) 50.0, 108.0 (123.35) 149.0
E 3616	84.0 (124.9) 152.0
E 1122	49.0 (75.93) 92.0
E 1121	35.0 (44.5) 56.0, 82.0 (103.42) 141.0

Figure 8 shows the bimodal distribution of the in situ spores isolated from *Palaeostachya elongata*.

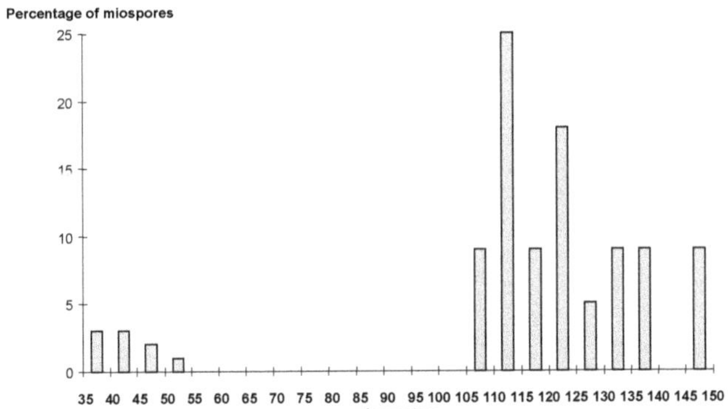

Figure 8. Size distribution of the in situ spores of *Palaeostachya elongate* (Presl) Weiss, (E 3631).

Palaeostachya gracillima Weiss 1876

Figure 7p–v

Material: No. E1119 is from the Kladno locality, Kladno-Rakovník Basin, Moscovian (Bolsovian), and No. E1127 is from the Příčina-Na Brantech locality near Lubná, Kladno-Rakovník Basin, Moscovian (Bolsovian).

A description of the in situ spores is as follows: Trilete spores are circular, subtriangular, or oval in amb. The laevigate exine of the microspores is up to 2 µm thick. The rays of the trilete mark reach a third to two-thirds of the radius. Secondary folds of the exine possess variable sizes, shapes, positions, and numbers (Figure 7q–v). The size range is 34 (65) 95 µm (Table 5). The labrum is 2–4 µm broad and is sometimes developed (Figure 7v). The in situ microspores can be compared to the dispersed miospore species *Calamospora braviradiata*, *C. minuta*, *C.* cf. *pedata*, and *C.* cf. *microrugosa*.

Table 5. Size of the in situ spores of *Palaeostachya gracillima* Weiss.

Specimen Number	Diameter of In Situ Spores (μm)
E 3635	48.0 (65.94) 83.0
E 1127	24.0 (62.57) 83.0
E 1119	42.0 (68.38) 95.0

Palaeostachya feistmantelii Němejc 1953
Figure 9a–f

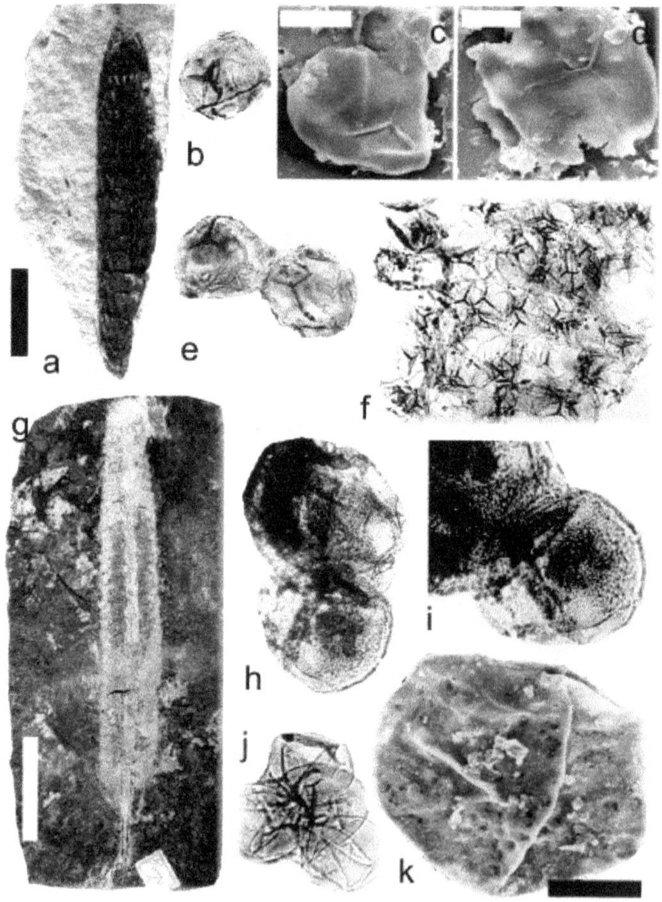

Figure 9. (**a**–**f**) *Palaeostachya feistmantelii* Němejc. (**a**) Specimen No. E3492. Štilec locality near Žebrák, Kladno-Rakovník Basin. Moscovian. Scale bar 20 mm. (**b**,**e**) In situ microspores of the *Calamospora* type. All ×500. (**c**,**d**) Fragments of in situ microspores of the *Calamospora* type. SEM, scale bar 20 μm. (**f**) Mass of damaged in situ microspores of the *Calamospora* type. ×500. (**g**–**k**) *Macrostachya carinata* (Germar) Zeiller. (**g**) Specimen No. E3637, Mirošov locality, Kladno-Rakovník Basin, Moscovian. Scale bar 50 mm. (**h**,**i**) In situ megaspores of the *Calamospora* type. All ×150. (**j**) In situ microspores of the *Calamospora* type. ×500. (**k**) In situ megaspore of the *Calamospora* type. SEM, scale bar 80 μm.

Macrostachya Schimper 1869.

Material: Nos. E3492 and E3493 are from the Štilec locality near Žebrák, Kladno, Rakovník Basin, Moscovian (Bolsovian).

A description of the in situ spores is as follows: The trilete are subcircular microspores. The size range is 55 (69) 108 μm. The laevigate exine is up to 2 μm thick. The labrum is 2–4 μm in size (Figure 9b,e). Rays of the trilete mark reach a third to two-thirds of the radius. The microspores are poorly preserved and only as fragments and incomplete specimens (Figure 9c,d,f), probably due to oxidation of the rock. This is the reason why in situ spores can be only described as the *Calamospora* type.

Figures 9g–k, 10 and 11a–f

Macrostachya carinata (Germar) Zeiller 1879

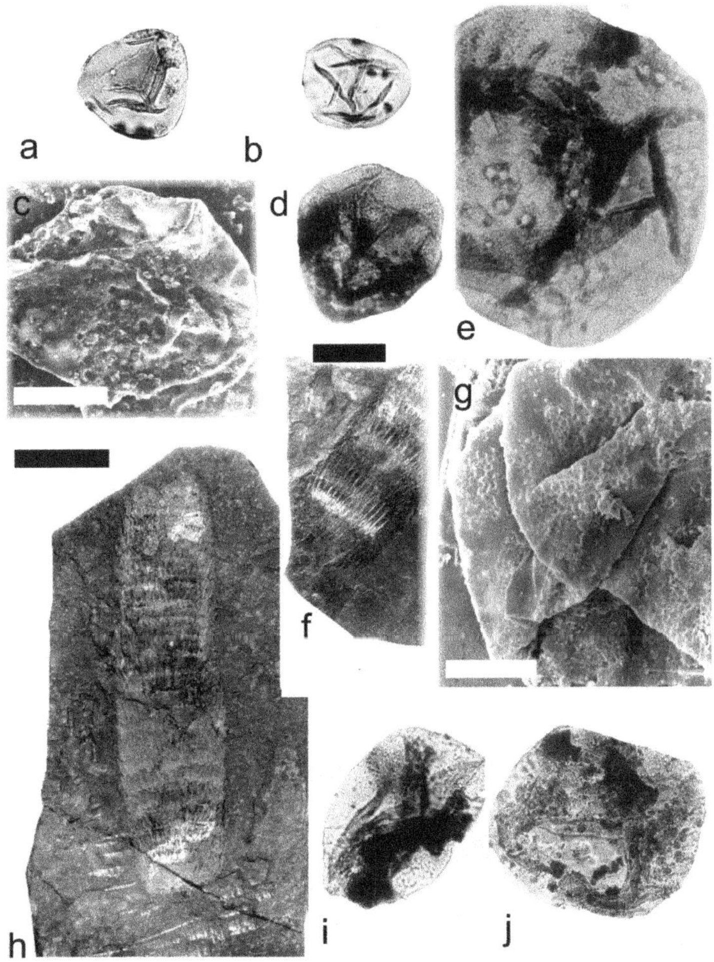

Figure 10. *Macrostachya carinata* (Germar) Zeiller. (**a–e**) Specimen No. E3637, Mirošov locality, Kladno-Rakovník Basin, Moscovian. (**a,b**) In situ microspores of the *Calamospora* type. All ×500. (**c**) In situ megaspore of the *Calamospora* type. SEM, scale bar 170 μm. (**d,e**) In situ megaspores of the *Calamospora* type. (**d**) ×120. Scale bar, (**e**) ×250. (**f,g**) Specimen No. E1180, Mirošov locality, Kladno-Rakovník Basin, Moscovian. (**f**) General view of the specimen. Scale bar 20 mm. (**g**) In situ megaspore of the *Calamospora* type. SEM. Note the negative sculpture of exine. Scale bar 700 μm. (**h–j**) Specimen No. E1181, Mirošov locality, Kladno-Rakovník Basin, Moscovian. (**h**) General view of the specimen. Scale bar 20 mm. (**i,j**) In situ megaspores of the *Calamospora* type. (**i**) ×120, (**j**) ×200.

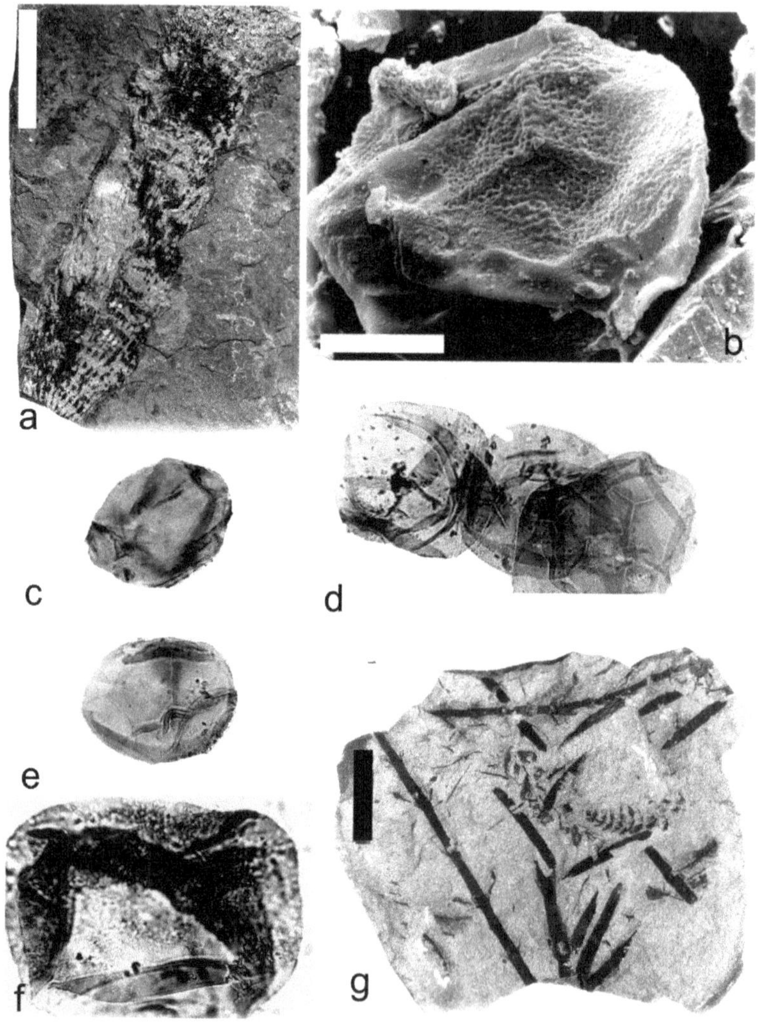

Figure 11. (**a**–**f**) *Macrostachya carinata* (Germar) Zeiller. (**a**) Specimen No E1178, Mirošov locality, Kladno-Rakovník Basin, Moscovian. Scale bar 40 mm. (**b**) In situ megaspore of the *Calamospora* type. Note the slightly elevated contact area and elevated labrum. SEM, scale bar 70 µm. (**c**–**e**) In situ microspores of the *Calamospora* type. All ×500. (**f**) In situ megaspore of the *Calamospora* type. ×150. (**g**) *Calamostachys germanica* Weiss, specimen E5641, Na Brantech locality, Lubná, Kladno-Rakovník Basin, Moscovian. Scale bar 40 mm.

Material: Collection of S. Opluštil, No. 861 from the Slaný Mine, Kladno-Rakovník Basin, Moscovian. No. E3641 is from the Mirošov locality, Kladno-Rakovník Basin, Moscovian (Asturian) E1178, E1181 (material type).

A description of the in situ spores is as follows: The trilete spores are circular, subtriangular, or oval in amb. The laevigate exine of the microspores (Figures 9j and 10a,b) is up to 2 µm and the megaspores (Figures 9h,i,k, 10c–e,i,j, and 11b–f) are up to 3 µm thick. The rays of the trilete mark reach a third to three-quarters of the radius. Secondary folds of the exine possess variable sizes, shapes, positions, and numbers (Figures 9j, 10a,b,e, and 11e,d). The size range (Table 6) of the microspores is 54 (76) 125 µm, and the megaspores are 156 (265)

372 μm (Table 7). The contact area and labrum are 2–4 μm in size and sometimes developed. The in situ microspores can be compared to the dispersed miospore species *Calamospora mutabilis*, *C.* cf. *pedata*, and *C.* cf. *liquida*. Megaspores are classified as *Calamospora* sp. due to their large diameters.

Table 6. Size of the in situ spores of *Macrostachya carinata* (Germar) Zeiller.

Specimens Number	Diameter of In Situ Spores (μm)
E 3541	66.0 (81.9) 102.0
E 3637 basal part	265.0 (318.57) 366.0
E 3637 middle part	221.0 (294.47) 372.0
E 3637 apical part	156.0 (203.2) 232.0
E 3637 apical part	58.0 (64.11) 69.0
E 1178	69.0 (119.67) 167.0
E 1180	183.0 (244.8) 342.0
E 1181	71.0 (93.0) 116.0

Table 7. Diameters of the in situ megaspores of the *Calamospora* type isolated from different portions (basal, middle, and apical) of the cone of *Macrostachya carinata* (Germar) Zeiller.

Number of Specimens	Diameter of Megaspores (μm)
E 3637 basal part	265.0 (318.57) 366.0
E 3637 middle part	221.0 (294.47) 372.0
E 3637 apical part	156.0 (203.2) 232.0

Calamostachys Schimper 1869

Calamostachys germanica Weiss 1976

Figures 11g, 12 and 13a–l

Material: Nos. E1161 and E2409 are from the Třemošná locality, Kladno-Rakovník Basin, Moscovian (Bolsovian); No. E2408 is from the Ignác Mine, Kladno-Rakovník Basin, Moscovian (Bolsovian); No. E3620 is from the Kladno locality, Kladno-Rakovník Basin, Moscovian (Bolsovian); No. E5641 is from the Na Brantech locality, Lubná, Kladno-Rakovník Basin, Moscovian (Bolsovian).

A description of the in situ spores is as follows: The trilete spores are circular, subtriangular, or oval in amb. The laevigate exine of the microspores is up to 2 μm thick. The rays of the trilete mark reach a quarter to a half of the radius. Secondary folds of the exine possess variable sizes, shapes, positions, and numbers (Figures 12a–h and 13c–j). The size range of the microspores is 54 (94) 126 μm, and the megaspores are 180 (357) 684 μm (Table 8). The labrum is 2–4 μm broad, and the dark contact areas (Figure 12a,d,f) are sometimes developed (Figures 12e and 13a). The in situ microspores can be compared to the dispersed miospore species *Calamospora microrugosa*, *C.* cf. *mutabilis*, *C. pallida*, *C.* cf. *pallida*, *C.* cf. *pedata*, *C. breviradiata*, and *C. hartungiana*; the megaspores are classified as *Calamospora* sp. due to their large diameters (up to 684 μm).

Table 8. Size of the in situ spores of *Calamostachys germanica* Weiss.

Specimen Number	Diameter of In Situ Spores (μm)
E 3620	75.0 (107.4) 126.0, 181.0 (247.83) 329.0
E 3621	261.0 (389.5) 485.0
E 2408	54.0 (61.27) 73.0
E 5641	55.0 (62.43) 91.0
E 1159	54.0 (79.9) 96.0
E 1161	180.0 (350.56) 438.0
E 2409	147.0 (442.05) 684.0

Figure 12. *Calamostachys germanica* Weiss. (**a–f**) Specimen E5641, Na Brantech locality, Lubná, Kladno-Rakovník Basin, Moscovian. (**a–f**) In situ microspores of the *Calamospora* type. Note the dark contact area (**a**,**c–f**). All ×500. (**g–j**) Specimen E5641, Třemošná locality, Lubná, Kladno-Rakovník Basin, Moscovian. (**g**) General view of the specimen. Scale bar 20 mm. (**h**,**j**) In situ megaspores of the *Calamospora* type. (**h**) ×150, (**j**) ×90. (**i**) In situ microspore of the *Calamospora* type. ×500. (**k**,**l**) Specimen No. E2409, Třemošná locality, Lubná, Kladno-Rakovník Basin, Moscovian. (**k**) General view of specimen. Scale bar 20 mm. (**l**) In situ megaspore of the *Calamospora* type. ×160. (**m**,**n**) Specimen No. E1159, Třemošná locality, Lubná, Kladno-Rakovník Basin, Moscovian. (**m**) General view of specimen. Scale bar 20 mm. (**n**) In situ microspore of the *Calamospora* type. Proximal surface. SEM, scale bar 20 μm.

Calamostachys incrassata Němejc 1953
Figures 13m and 14a–g

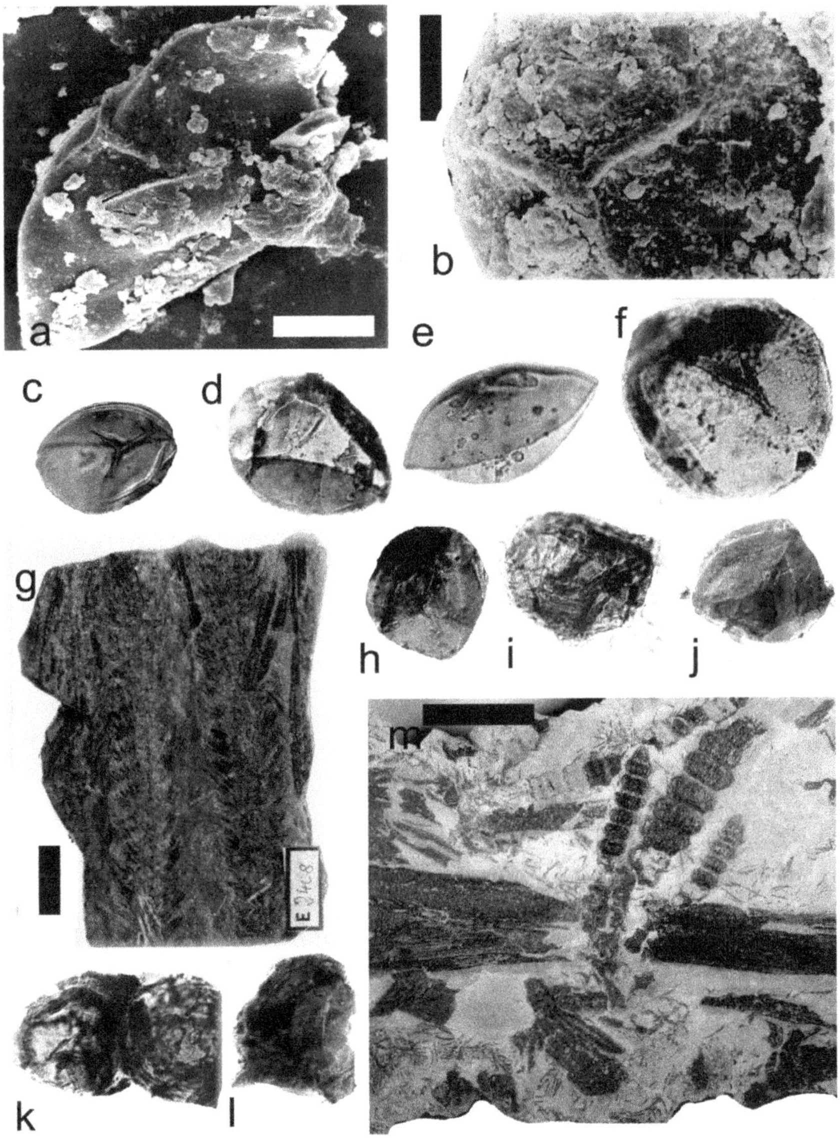

Figure 13. (**a–k**) *Calamostachys germanica* Weiss. (**a–f**) Specimen No. E1159, Třemošná locality, Lubná, Kladno-Rakovník Basin, Moscovian. Scale bar. (**a**) In situ microspore of the *Calamospora* type. SEM, scale bar 20 µm. (**b**) In situ megaspore of the *Calamospora* type. SEM, scale bar 40 µm. (**c–e**) In situ microspores of the *Calamospora* type. ×500. (**f**) In situ megaspore of the *Calamospora* type. Scale bar. (**g–l**) Specimen No. E2408, Ignác Mine, Kladno-Rakovník Basin, Moscovian. (**g**) General view of the specimens. Scale bar 15 mm. (**h–l**) In situ microspores of the *Calamospora* type. Note the irregular fragments of the exospore. All ×500. (**m**) *Calamostachys incrassata* Němejc. Specimen No. E. E1114, V Krčeláku locality, Rako Mine near Lubná, Kladno-Rakovník Basin, Moscovian. Scale bar 50 mm.

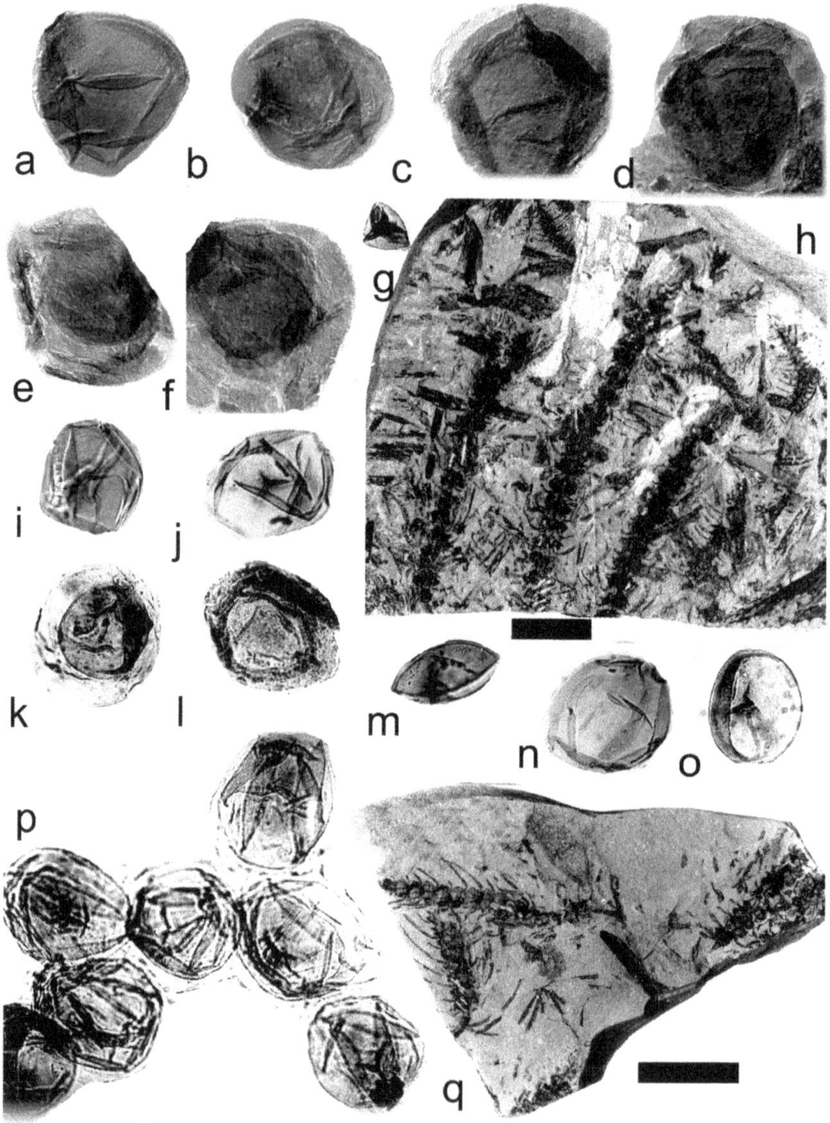

Figure 14. (**a–g**) *Calamostachys incrassata* Němejc, specimen No. E. E1114, V Krčeláku locality, Rako Mine near Lubná, Kladno-Rakovník Basin, Moscovian. (**a–f**) In situ microspores of the *Calamospora* type. Note the fragments of probably tapetal tissues. All ×500. (**g**) Immature in situ microspore. ×500. (**h–q**) *Calamostachys longibracteata* Němejc. (**h–p**) Specimen No. E1163, Mayrau Mine, Vinařice, Kladno-Rakovník Basin, Moscovian. (**h**) General view of the specimen. Scale bar 20 mm. (**i,j,n,o**) In situ microspores of the *Calamospora* type. All ×500. (**k**) In situ microspore of the *Calamospora* type. Note that monopseudosaccate-like layer envelopes the inner body of the *Calamospora* type. ×500. (**l**) In situ microspore of the *Calamospora* type. Note the monopseudosaccate-like layer. ×500. (**m**) In situ microspore of the *Calamospora* type. Note that one major fold covers the microspore body. ×500. (**p**) In situ microspores of the *Calamospora* type. Note that monopseudosaccate-like exine layers enveloped central bodies. ×450. (**q**) Specimen No. E1155, Max Mine, Libušín, Kladno-Rakovník Basin, Bolsovian. Scale bar 30 mm.

Material: No. E1114, is from the V Krčeláku locality, Rako Mine near the Lubná, Kladno-Rakovník Basin, Moscovian (Bolsovian).

A description of the in situ spores is as follows: The trilete spores are circular, subtriangular, or oval in amb. The laevigate exine of the microspores is up to 2 μm thick. The rays of the trilete mark reach a third of the radius. Secondary folds of the exine possess variable sizes, shapes, positions, and numbers (Figure 14a–f). The size range is 51 (73) 88 μm. The labrum is 2–4 μm broad and is sometimes developed. The in situ microspores can be compared to the dispersed miospore species *Calamospora* cf. *hartungiana*. Some spores are 54 (75) 94 μm large and are enveloped in a very thin monopseudosaccate layer (Figure 14c–f) and may resemble some forms of the miospores genera *Auroraspora* Hoffmeister et al., *Remysporites* Butterworth & Williams, *Perotrilites* Couper, *Phyllothecotriletes* Luber, or even *Diaphanospora* Balme.

Calamostachys longibracteata Němejc 1953

Figures 14h–q and 15a–m

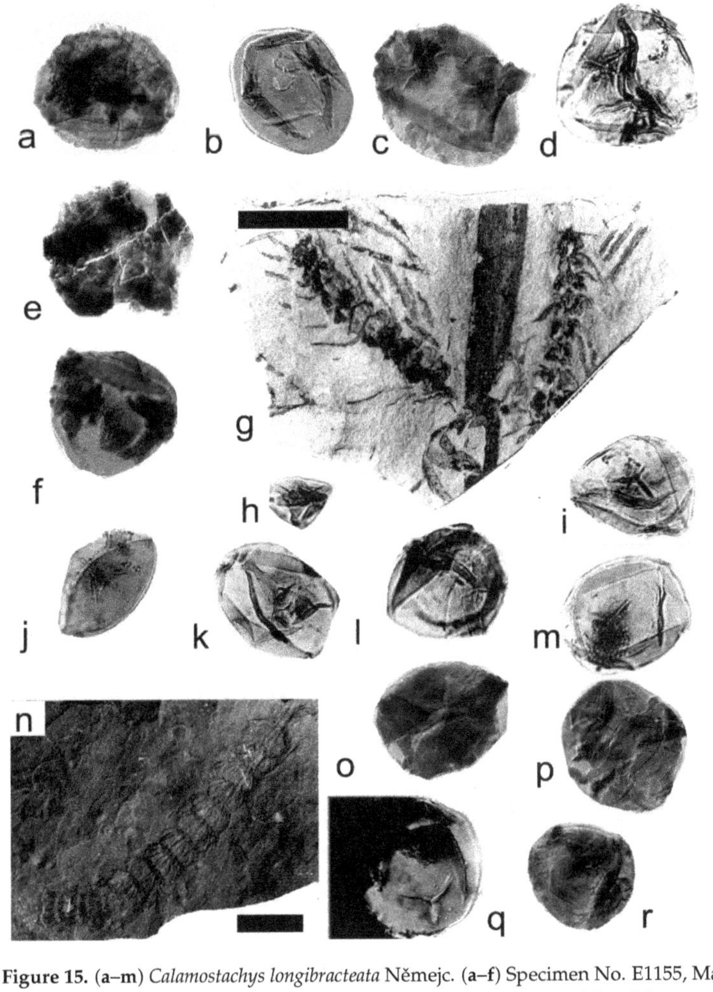

Figure 15. (**a–m**) *Calamostachys longibracteata* Němejc. (**a–f**) Specimen No. E1155, Max Mine, Libušín, Kladno-Rakovník Basin, Moscovian. (**a,c,e,f**) In situ microspores of the *Calamospora* type. Note the fragments of the monopseudosaccate-like exine layer. All ×500. (**b,d**) In situ microspores of the *Calamospora* type. All ×500. (**g–m**) Specimen E1154. Max Mine, Libušín, Kladno-Rakovník Basin. Moscovian. General view. (**g**) General view of the specimen. Scale bar 20 mm. (**h**) Immature in situ

microspore. ×500. (**i**) In situ microspore of the *Calamospora* type. ×500. (**j**) In situ microspore of the *Calamospora* type. Note that one major fold of exine covers the body. ×500. (**k–m**) In situ microspores of the *Calamospora* type. All ×500. (**n–r**) *Calamostachys tuberculata* Sternberg, E 1147, Doubrava locality, Kladno-Rakovník Basin, Moscovian. General view. Scale bar. (**n**) General view of specimen. Scale bar 20 mm. (**o,p,r**) In situ microspore of the *Calamospora* type. Note the fragments of probable tapetal tissues. All ×500. (**q**) In situ microspores of the *Calamospora* type. ×500.

Material: No. E3605 is from the Kladno locality, Kladno-Rakovník Basin, Moscovian (Bolsovian); Nos. E1154 and E1155 are from the Max Mine, Libušín, Kladno-Rakovník Basin, Moscovian (Bolsovian); No. E1163 is from the Mayrau Mine, Vinařice, Kladno-Rakovník Basin, Moscovian (Bolsovian).

A description of the in situ spores is as follows: The trilete spores are circular, subtriangular, or oval in amb. The laevigate exine of the microspores is up to 2 μm thick. The rays of the trilete mark reach a third to two-thirds of the radius. Secondary folds of the exine possess variable sizes, shapes, positions, and numbers (Figure 14a–c,i,j,m–o). The size range is 45 (60) 83 μm (Table 9). Sometimes, the outer perispore-like layer envelopes the central body of the *Calamospora* type (Figure 14k,l,p). The in situ microspores can be compared to the dispersed miospore species *Calamospora microrugosa*, *C. pallida*, and *C. pedata*.

Table 9. Size of the in situ spores of *Calamostachys longibracteata* Němejc.

Specimen Number	Diameter of In Situ Spores (μm)
E 3605	45.0 (54.5) 72.0
E 1154	45.0 (63.33) 78.0
E 1155	48.0 (58.5) 77.0
E 1163	47.0 (62.21) 84.0

Calamostachys tuberculata (Sternberg) Jongmans 1911

Figures 15n–r and 16

Material: Collection of S. Opluštil: Nos. 1238 and 1951 are from the Kladno locality, Kladno-Rakovník Basin, Kasimovian (Lower Stephanian); No. E3589 is from the Lubná locality, Kladno-Rakovník Basin, Moscovian (Bolsovian); E1152 is from the Mirošov locality, Kladno-Rakovnk Basin, Moscovian (Asturian); No. E1148 is from the Doubrava locality, Kladno-Rakovník Basin, Moscovian (Asturian); No. E1147 is from the Kladno locality, Kladno-Rakovník Basin, Moscovian (Bolsovian).

A description of the in situ spores is as follows: The trilete spores are circular, subtriangular, or oval in amb. The laevigate exine of the microspores is up to 2 μm thick. The rays of the trilete mark reach a third to two-thirds of the radius. Secondary folds of the exine possess variable sizes, shapes, positions, and numbers (Figures 15q and 16c,d). The size range is 30 (61) 110 μm in diameter (Table 10). The labrum is 2–4 μm broad and is sometimes developed. Some specimens have irregular fragments of perisporial tissue (Figures 15o,p,r and 16a–d). The in situ microspores can be compared to the dispersed miospore species *Calamospora microrugosa*, *C.* cf. *pedata*, *C. straminea*, and *C. breviradiata*.

Table 10. Size of the in situ spores of *Calamostachys tuberculata* (Sternberg) Jongmans.

Specimen Number	Diameter of In Situ Spores (μm)
E 3589	318.0 (352.67) 396.0
E 1147 basal part	56.0 (67.36) 74.0
apical part	54.0 (64.3) 75.0
E 1148	51.0 (65.21) 78.0
E 1152	52.0 (65.9) 79.0
1238	31.0 (66.23) 110.0
1951	30.0 (62.6) 107.0

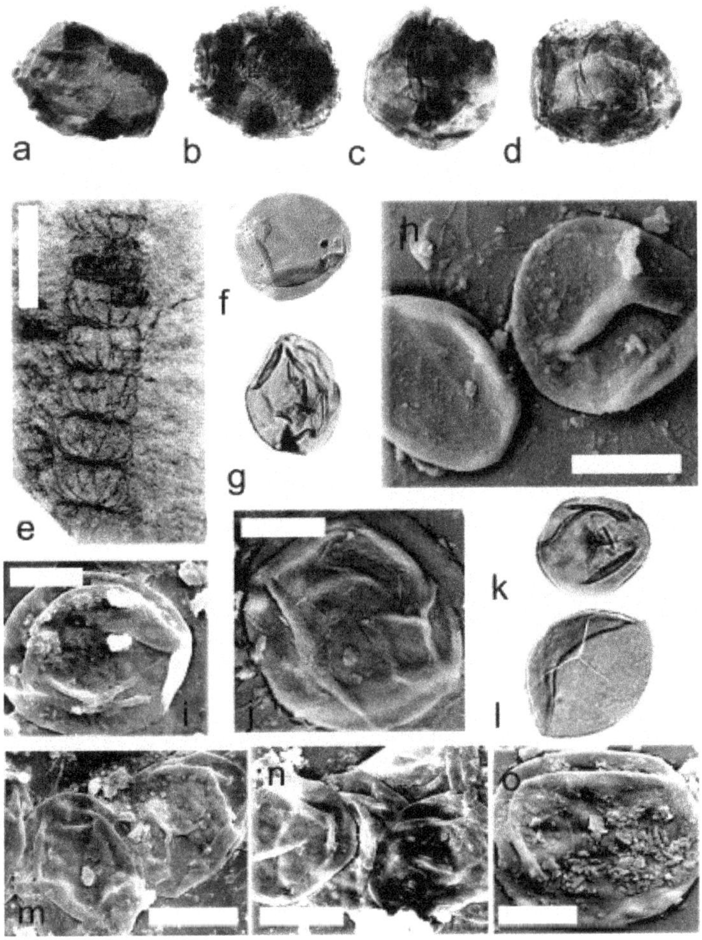

Figure 16. *Calamospora tuberculata* Sternberg. (**a–d**) Specimens E 1147, Doubrava locality, Kladno-Rakovník Basin, Moscovian. (**a–d**) In situ microspores of the *Calamospora* type. Note the irregular fragments of probable tapetal tissues. All ×500. (**e–o**) Specimen 1951, Kladno locality, Kladno-Rakovník Basin, Kasimovian. (**e**) General view of the specimen. Scale bar 15 mm. (**f**,**g**,**k**,**l**) In situ microspores of the *Calamospora* type. All ×500. (**h–j**,**m–o**) In situ microspores of the *Calamospora* type. SEM. (**h**) Scale bar 40 µm. (**i**) Scale bar 35 µm. (**j**) Scale bar 30 µm. (**m**) Scale bar 40 µm. (**n**) Scale bar 40 µm. (**o**) Scale bar 25 µm.

Calamostachys cf. *ramosa* Weiss 1884

Figure 17a–l

Material: No. E3634 is from the Maria-Julia Mine, Žacléř, Intra-Sudetic Basin, Bashkirian (Langsettian-Duckmantian); No. E3627 is from the Ronna Mine, Kladno-Rakovník Basin. Moscovian (Bolsovian).

A description of the in situ spores is as follows: The trilete spores are circular, subtriangular, or oval in amb. The laevigate exine of the microspores is up to 2 µm thick. The rays of the trilete mark reach a third of the radius. Secondary folds of the exine possess variable sizes, shapes, positions, and numbers (Figure 17c,d,h,j,l). The size range is 48 (62) 89 µm in diameter. The in situ microspores can be compared to the dispersed miospore species *Calamospora* cf. *microrugosa*. Some specimens are 49 (65) 99 µm in size and are enveloped by a thin monopseudosaccate layer and can resemble some forms of the miospore genera *Auroraspora*, *Remysporites*, *Perotrilites*, *Phyllothecotriletes*, or even *Diaphanospora*, and some others have irregular fragments of perisporial tissue (Figure 17e,h,i,k,l).

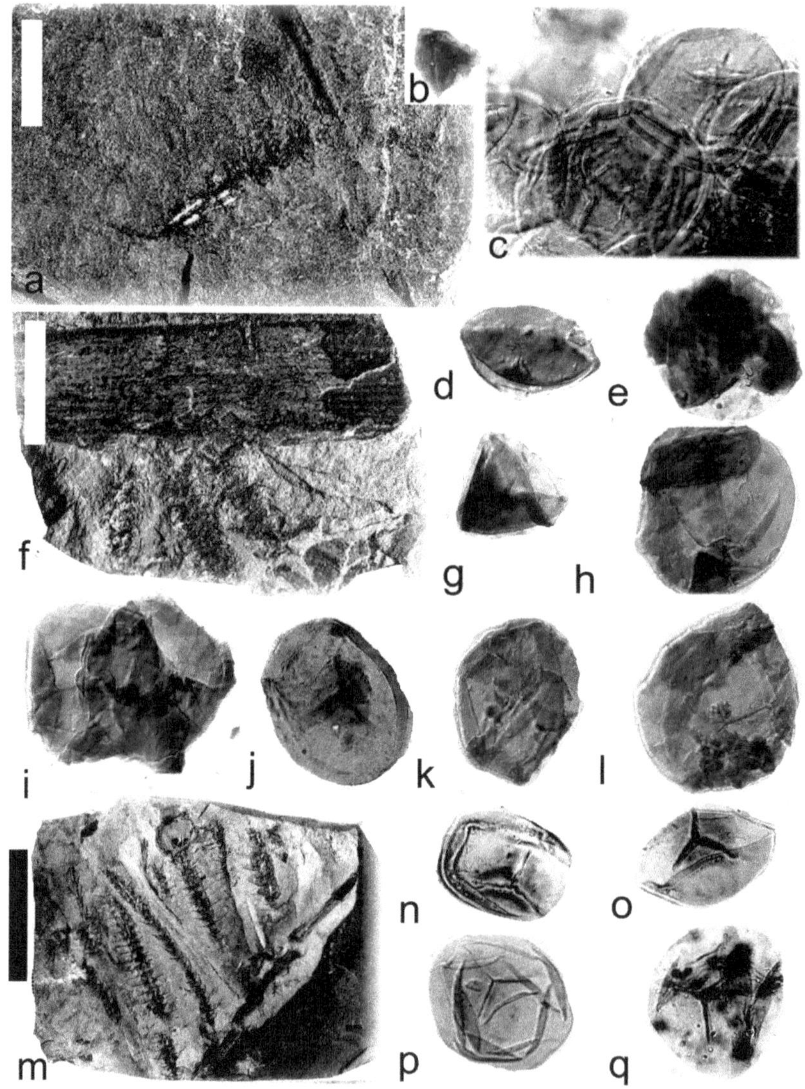

Figure 17. (**a–l**) *Calamostachys* cf. *ramosa* Weiss, specimen E3627, Ronna Mine, Kladno-Rakovník Basin, Moscovian. (**a**) General view of specimens. Scale bar 15 mm. (**b**) Immature in situ microspore. ×500. (**c**) In situ microspores of the *Calamospora* type. ×500. (**d**) In situ microspore of the *Calamospora* type. Note one major fold covering the body. ×500. (**e**) In situ microspore of the *Calamospora* type with fragments of probable tapetal tissues. ×500. (**f–l**) Specimen E E3634, Maria-Julia Mine, Žacléř, Intra-Sudetic Basin. (**f**) General view of the specimen. Scale bar 15 mm. (**g**) Immature in situ microspore. ×500. (**h–l**) In situ microspores of the *Calamospora* type. Note the fragments of probable tapetal tissues. (**j**) Note the dark contact area. All ×500. (**m–q**) *Calamostachys intermedia* Feistmantel. Specimen E2410, Stradonice locality near Beroun, Kladno-Rakovník Basin. (**m**) General view of the specimen. Scale bar 40 mm. (**n–q**) In situ microspores of the *Calamospora* type. All ×500.

Calamostachys intermedia Feistmantel 1872

Figure 17m–q

Material: Nos. E2410 and E1174 are from the Stradonice locality near Beroun, Kladno-Rakovník Basin, Moscovian (Bolsovian).

A description of the in situ spores is as follows: The trilete spores are circular, subtriangular, or oval in amb. The laevigate exine of the microspores is up to 2 µm thick. The rays of the trilete mark reach half to two-thirds of the radius. Secondary folds of the exine possess variable sizes, shapes, positions, and numbers (Figure 17n–q). The size range is 34 (85) 137 µm. The labrum is 2–4 µm broad and is sometimes developed (Figure 17n,o). The in situ microspores can be compared to the dispersed miospore species *Calamospora* cf. *liquida* and C. cf. *pedata*.

Calamostachys grandis (Zeiller) Jongmans 1911

Figure 18a–f

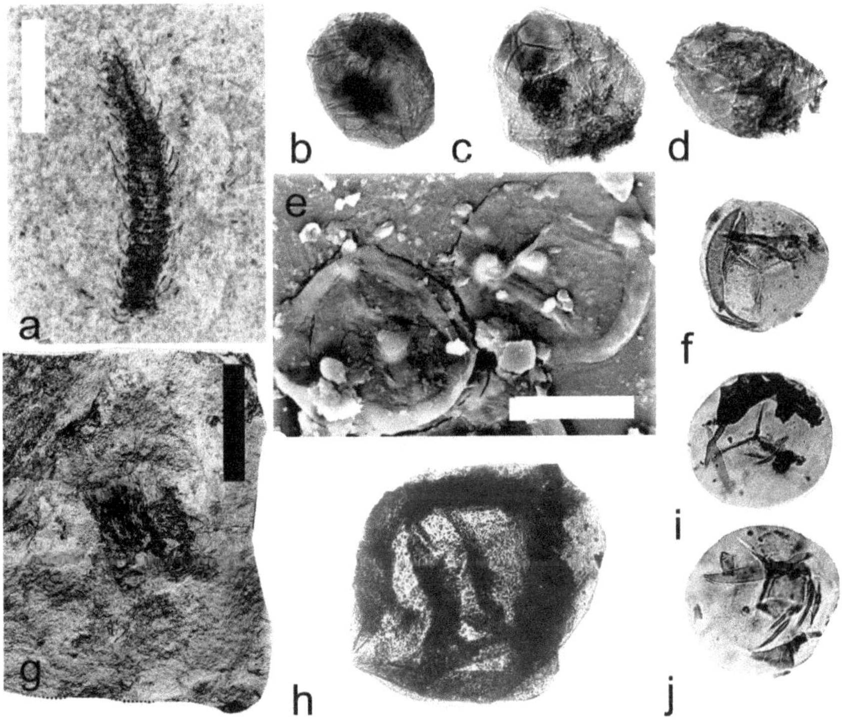

Figure 18. (**a**–**f**) *Calamostachys grandis* (Zeiller) Jongmans, specimen 1625, Kladno locality, Kladno-Rakovník Basin, Moscovian. (**a**) General view of the specimen. Scale bar 12 mm. (**b**–**d**) In situ microspores of the *Calamospora* type. Note the fragments of probable tapetal tissues. All ×500. (**e**) In situ microspores of the *Calamospora* type. SEM, scale bar 30 µm. (**f**) In situ microspore of the *Calamospora* type. ×500. (**g**–**j**) *Calamostachys* sp. E3639, Tuchlovice locality, Kladno-Rakovník Basin, Moscovian. (**g**) General view of the specimen. Scale bar 25 mm. (**h**) In situ megaspore of the *Calamospora* type. ×150. (**i**,**j**) In situ microspores of the *Calamospora* type. All ×500. *Calamostachys* sp.

Material: Collection of S. Opluštil, No. 1625, from the Kladno locality, Kladno-Rakovník Basin, Moscovian (Bolsovian).

A description of the in situ spores is as follows: The trilete spores are circular, subtriangular, or oval in amb. The laevigate exine of the microspores is up to 2 µm thick. The rays of the trilete mark reach a third to a half of the radius. Secondary folds of the exine possess variable sizes, shapes, positions, and numbers (Figure 18b–f). The size range is 52

(58) 70 µm. The in situ microspores can be compared to the dispersed miospore species *Calamospora breviradiata*.

Figure 18g–j

Material: Nos. E3638 and E3639 are from the Tuchlovice locality, Kladno-Rakovník Basin, Moscovian (Bolsovian).

A description of the in situ spores is as follows: The trilete spores are circular, subtriangular, or oval in amb. The laevigate exine of the microspores is up to 2 µm thick. The rays of the trilete mark reach three-quarters of the radius. Secondary folds of the exine possess variable sizes, shapes, positions, and numbers (Figure 18i,j). The size range of the microspores is 45 (67) 82 µm, and the megaspores are (Figure 18h) 610 (703) 815 µm. The labrum is 2–4 µm broad and is sometimes developed. The in situ microspores can be compared to the dispersed miospore species *Calamospora* cf. *liquida* and *C.* cf. *flexilis*, and the megaspores are classified as *Calamospora* sp. due to their large diameters.

Huttonia Sternberg 1837

Huttonia spicata Sternberg 1837

Figure 19

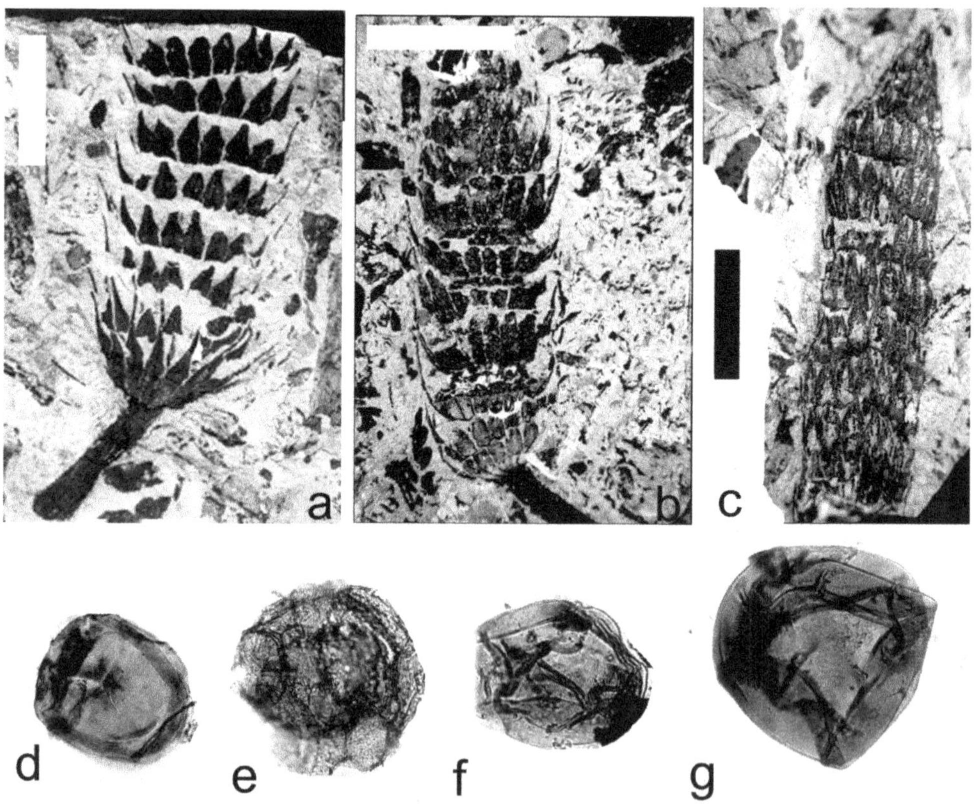

Figure 19. *Huttonia spicata* Sternberg. (**a**) Specimen E74, Vranovice locality. Radnice Basin, Moscovian, Scale bar 20 mm. (**b**) Specimen E2419. Vranovice locality. Radnice Basin, Moscovian. Scale bar 20 mm. (**c**) Specimen E75. Vranovice locality. Radnice Basin, Moscovian. Scale bar 20 mm. (**d**) In situ microspore of the *Calamospora* type. ×500. (**e**) In situ microspores of the *Calamospora* type. Note the outer exine monopseudosaccate-like layer. ×500. (**f**) In situ microspore of the *Calamospora* type. ×500. (**g**) In situ megaspore of the *Calamospora* type. ×120.

Material: No. E3614 is from the Ovčín locality, Radnice Basin, Moscovian (Bolsovian); Nos. E2419, E75, and E76 are from the Vranovice locality, Kladno-Rakovník Basin, Moscovian (Bolsovian).

A description of the in situ spores is as follows: The trilete spores are circular, subtriangular, or oval in amb. The laevigate exine of the microspores is up to 1–4 μm thick. The rays of the trilete mark reach a third to two-thirds of the radius. Secondary folds of the exine possess variable sizes, shapes, positions, and numbers (Figure 19d,f,g). The size range of the microspores is 66 (84) 114 μm, and the megaspores are (Figure 19g) 115 (166) 240 μm. The labrum is 2–4 μm broad and is sometimes developed. The in situ microspores can be compared to the dispersed miospore species *Calamospora* cf. *breviradiata*, *C.* cf. *pedata*, and the megaspores are of the *Calamospora laevigata* type. Some specimens are enveloped in a thin monopseudosacccate exine layer (Figure 18e) and can resemble some forms of miospore genera *Auroraspora*, *Remysporites*, *Perotrilites*, or even *Diaphanospora*.

5. *Calamospora* Spores

5.1. Morphological Criteria

The majority of morphological features are important for the classification of dispersed calamospores, but some others, like the thickness of the exine and the sculpture, are not as significant. Some species are morphologically closely similar, and it is possible that they can be synonymous. We can divide the dispersed calamospores into a few main morphological groups, e.g., by contact area, by contact area and labrum, and only by the labrum.

Only a few authors have described a significant number of calamitean cones and their in situ *Calamospora* micro- and megaspores [11,15,24].

5.1.1. Diameter

The size range given for every dispersed *Calamospora* species is not a significant criterion for their natural classification. In every *Calamospora* in situ population, the size range is variable, with specimens ranging from about 20–30 μm to more than 100 μm. One of the morphological species groups shares the same features and may be the only criterion for distinguishing it among different species of dispersed spore species [11]. The average size range of the in situ *Calamospora* microspores is 24 (75.02) 181 μm, and 147.0 (399.15) 815.0 μm for the *Calamospora* megaspores. The average difference (i.e., among the smallest and largest spores in a slide) in size is 40 μm. Dispersed calamospores are assigned to miospore and megaspore species based on their size, but the arbitrary size criterion of at least 200 μm for megaspores need not always correspond with their biological function. For example, the spore population from *Palaeostachya elongata* (E3631) is divided into two size groups. The smaller group ranges in size from 35 to 55 μm, and the larger group ranges from 105 to 150 μm in diameter. Similarly, in another specimen of *P. elongata* (E1121), spores fall into two size categories: the first category includes spores 35 to 55 μm in diameter, and the second group includes spores ranging from 85 to 140 μm in diameter. The definition of 200 μm or more for megaspores is artificial and may not apply to many *Calamospora* megaspores. The diameter of calamospores can be influenced by stages of maturity. It seems that calamitean cones matured gradually from apical to basal sporangia. The difference in the diameter for *Calamospora* from apical and basal sporangia is 5 (12) 16 μm, with an extreme example of cones of *Palaeostachya elongata*, where it is 79 μm. The difference between relatively immature and mature specimens was about 91 μm on average for the megaspores.

5.1.2. Contact Area

Usually, all spores in a *Calamospora* in situ population either do or do not have a contact area. This could mean that the occurrence of the contact area may be a reliable criterion for the classification of dispersed calamospores. However, the number of in situ microspores with a contact area is higher in apical than in basal microsporangia, i.e., it implies that this morphological feature may be related to different degrees of maturity because the contact area means that the exine at the proximal pole is thickened and the thickness of the exine may be influenced by stages of maturity. The diameter of the contact area is usually equal to the length of the rays of the trilete marks or maybe sometimes slightly shorter. Sometimes, under SEM, the contact area can be seen to be slightly elevated, i.e., it is slightly thickened proximally. We know (minimally) of ten Pennsylvanian *Calamospora* species with contact areas with a size range from 34 to 146 μm and with the length of the trilete marks being a third to two-thirds of the radius. This indicates that the size of the contact area is not an easy way to distinguish among *Calamospora* dispersed species.

5.1.3. Length of Rays of the Trilete Mark

The length of the rays of the trilete mark of a majority of *Calamospora* specimens is from a third to three-fourths of the radius. Almost all dispersed calamospores fit into this size range. Only a few dispersed *Calamospora* species have longer rays of the trilete mark, e.g., *Calamospora liquida*.

5.1.4. Labrum

The presence or absence of a labrum need not be a constant morphological feature. Some spores from one sporangium possess a labrum, but some other spores from the same sporangium do not. A labrum is usually seen well using SEM. Only a few dispersed *Calamospora* species are defined with a labrum, e.g., *C. flava* and *C. elliptica*.

5.1.5. Secondary Folds of Exine

We can imagine *Calamospora* spores in sporangium like small circular balloons without any folding, only with trilete marks on the exine surface. Due to the fossilization (adpression specimens, not petrifactions), the spore body is compressed, and the thin laevigate exine becomes folded. Different types of folding are purely occasional and not biological. Many dispersed *Calamospora* species are distinguished based on different foldings. It shows the arbitrary character of the classification of dispersed *Calamospora* species. For example, *C. pedata* is typified by one major secondary fold that covers about a half of the spore body; *C. flexilis* is typical by its folds parallel with the rays of the trilete mark, and *C. mutabilis*, *C. straminea*, *C. parva*, and *C. breviradiata* have folds parallel with the margins of the spores.

5.1.6. Fragments of Tapetal Tissues

Sometimes, it is possible to observe calamospores with various fragments of probable tapetal tissues on the exine surface. These fragments have irregular sizes, shapes, thicknesss, and numbers. Sometimes, they cover the majority of the exine surface, and sometimes, only a small part. These forms are not comparable to any dispersed miospore species and are more likely related to the ontogenetic stages.

Another feature associated with the ontogenetic stages is a delicate circular monopseudosaccate-like layer that envelopes spores of the *Calamospora* type. These forms were described in [25] for the *Calamostachys calathifera* [26], *C. williamsoniana* [27], *Palaeostachya feistmanteli*, *Calamostachys calathifera*, *C. binneyana*, *C. germanica* [28,29], *C. incrassata*, and *Huttonia spicata*. Here, these forms were observed in spore populations isolated from *Palaeostachya feistmanteli*, *P. distachya*, *Calamostachys longibracteata*, *C. carinata* and *C.* cf. *ramosa*, and *Huttonia spicata*. Some authors [26,27] compared these forms with some species of miospore genera, such as *Auroraspora*, *Perotrilites*, *Remysporites*, *Callialasporites*, *Phyllothecotriletes*, or even *Diaphanospora*. But they are immature forms, i.e., younger ontogenetic stages, and are not different spore taxa. Fully matured spores lack a monopseudosaccus-like layer.

6. Elaterites

Spores of the genus *Elaterites* Wilson are very rare in the dispersed record. The genus was established [30] for spores with three elaters enveloping a central body of the *Calamospora* type. These microspores are very rarely reported [30–33]. Seventeen species of genera, such as the *Calamocarpon* Baxter, *Calamostachys*, *Mazostachys*, *Palaeostachya*, *Pendulostachys* Good, *Pothocites* Paterson, and *Weissistachys* Rothwell & Taylor, yielded spores of the *Elaterites* type from mainly petrified (Table 11) [9,21] and adpression [15] specimens. The size range of these microspores is 38 (78) 280 μm, and almost all of them are 38–112 μm in diameter, except for those isolated from *Calamostachys americana* Arnold, which is unusually large (140–280 μm) [10]. Some authors [10,12,34] propose that all calamiteans produced microspores of the *Elaterites* type. We have studied hundreds of palynological slides with in situ *Calamospora* populations in different stages of ontogeny, macerated from sixty specimens of calamitean cones, and we have never seen any elaters or any elater-like structures or their fragments. It is evident that there is a group of calamitean cones that produced spores of the *Elaterites* type, but the majority of them yielded only spores of the *Calamospora* type. Some authors [10,12,24] have proposed that *Vestispora* represents an ontogenetic stage of *Calamospora*. The authors of [35] excluded this theory based on some main points, and the authors confirmed their conclusions.

Table 11. Parent plants produced spores of the Elaterites type.

Parent Plant	Diameter of Microspores (μm)	Classification of In Situ Spores	References
Calamocarpon insigne	38–60	*Calamospora*/*Elaterites*	[24] Good 1975
Calamostachys americana	140–280	*Calamospora*/*Elaterites*	[24] Good 1975
C. binneyana	38–61	*Calamospora*/*Elaterites*	[24] Good 1975
C. casheana	75	*Calamospora*/*Elaterites*	[24] Good 1975
C. inversibractus	44–74	*Calamospora*/*Elaterites*	[24] Good 1975
C. ludwigi	74–112	*Calamospora*/*Elaterites?*	[15] Hartung 1933
C. magnae-crucis	45	*Calamospora*/*Elaterites?*	[7] Balme 1995
Mazostachys noei	45–50	*Calamospora*/*Elaterites*	[24] Good 1975
M. pedunculata	52–100	*Calamospora*/*Elaterites*	[24] Good 1975
Palaeostachya andrewsii	56–100	*Calamospora*/*Elaterites*	[24] Good 1975
P. decacnema	54	*Calamospora*/*Elaterites*	[24] Good 1975
P. distachya	70–105	*Calamospora*/*Elaterites?*	[24] Good 1975
P. feistmanteli	45–95	*Calamospora*/*Elaterites?*	[26] Serret Brousmiche 1986
P. vera	75–80	*Calamospora*/*Elaterites*	[24] Good 1975
Pendulostachys cingulariformis	63–89	*Calamospora*/*Elaterites*	[24] Good 1975
Pothocites grantoni	82–104	*Calamospora*/*Elaterites?*	[7] Balme 1995
Weissistachys kentuckiensis	39–75	*Calamospora*/*Elaterites*	[24] Good 1975

Vestispora is characterized by a homogeneous outermost exine layer and small circular operculum lying above the proximal pole of the central body of the *Calamospora* type. *Elaterite* spores lack an operculum, and there is always a space among three elaters, i.e., the outer layer is not homogeneous. Elaters of *Elaterites* originate from a small triangular area on the distal pole, i.e., opposite to the proximal surface with a small circular operculum. Anything resembling such a triangular distal structure has never been observed on the *Calamospora* spores described herein. The structure of *Elaterites* is monotonous, but the sculpture of *Vestispora* is variable, including laevigate, foveolate, costate, and primary and secondary reticulate.

The second point is that *Vestispora* and *Calamospora* spores were produced by different plants. *Vestispora* was not produced by any calamiteans. The major *Vestispora* producer was one group of sphenophyllaleans [33]. Another small group of sphenophyllaleans produced calamospores but only *Calamospora* and not *Calamospora* and *Vestispora* together. *Calamospora* ranges from the Devonian to the Cenozoic, whereas *Vestispora*-producers range from the Brigantian to late Kasimovian. As a consequence, almost all *Vestispora*-producing sphenophyllaleans are good stratigraphical markers, but calamitean spores are not. *Elaterites* have an even shorter stratigraphical range, ranging through only a part of the Pennsylvanian [36]. Almost all *Vestispora*-producing sphenophyllaleans are good stratigraphical markers [37], but calamiteans are not. *Elaterites* have an even shorter stratigraphical range, only a part of the Pennsylvanian [35].

The third point concerns different ultrastructure sections of the exine (TEM) of *Elaterites* [38] and *Vestispora* [39]. Elaters of *Elaterites* are three-layered [39], while the exospore of *Vestispora* is only bi-layered [40].

When in situ, *Elaterites* are reported only from coal-ball calamitean cones and not from adpressions. The only roughly similar spore structures are hygroscopic elaters of the recent *Equisetum* spores, which show that their ultrastructure is different. The sporoderm of *Equisetum* spores has an endospore, exospore, perispore, and bi-layered elaters [38], but the *Elaterites* are only bi-layered, and elaters have three layers.

Vestispora spores have a circular laevigate central body with trilete marks and outer exospores with a circular operculum. Sometimes, in palynological slides with in situ *Vestispora*, it is possible to observe various degrees of mechanical damage in the exospores and that the central body is of the *Calamospora* type. However, *Vestispora* is not a relatively immature *Calamospora* because we have never observed any ontogenetic stages of *Vestispora*. Table 11 shows all the *Elaterites* producing calamitean plants.

7. Parent Plants

The main *Calamospora* producers are calamitean cones, especially the genera *Calamostachys*, *Palaeostachya*, and *Macrostachya*. Minor producers of *Calamospora* are the genera Pothocites, *Paracalamostachys* Weiss, *Huttonia*, *Weissistachys*, *Cingularia* Weiss, *Pendulostachys*, and *Calamocarpon*. Non-calamitean producers are sphenophyllaleans (Table 12) and noeggerathialeans. Sphenophyllaleans and calamiteans are closely related. The noeggerathialean genus *Discinites* K. Feistmantel produced micro- and megaspores of the *Calamospora* type and *Noeggerathiaestrobus* O. Feistmantel-only *Calamospora* megaspores (Table 12).

Table 12. Paleozoic spores of Equisetales and their parent plants.

Spore Genus	Calamiteans
Calamospora (micro- and megaspores)	*Pothocites grantonii*, *P. pettycurensis*, *Calamostachys americana*, *C. Binneyana*, *C. calathifera*, *C. casheana*, *C. dumasi*, *C. germanica*, *C. grandis*, *C. gunlongii*, *C. incrassata*, *C. intermedia*, *C. inversibractus*, *C. longibraceata*, *C. ludwigi*, *C. magnae-crucis*, *C. paeniculata*, *C.* cf. *ramosa*, *C. solmsii*, *C. tuberculata*, *C.* cf. *tuberculata*, *C. williamsoniana*, *C. zeilleri*, *C.* sp (*sensu* Moore), *C.* sp. A (*sensu* Bek), *C.* sp. B (*sensu* Serret and Brousmiche), *Macrostachya carinata*, *M. carinata* var. *approximata*, *M. caudata*, *M. hauchecornei*, *M. infundibuliformis*, *M. thompsonii*, *Palaeostachya andrewsii*, *P. aperta*, *P. decacnema*, *P. dircei*, *P. distachya*, *P. elongata*, *P. ettingshausenii*, *P. equisetoformis*, *P. feistmantelii*, *P. gracilis*, *P. gracillima*, *P. pedunculata*, *P. superba*, *P. thuringiaca*, *P. trabeculata*, *P. vera*, *P* sp. (*sensu* Moore), *P.* sp. A (*sensu* Bek), *Paracalamostachys scatervillei*, *P. heterospora*, *P. minor*, *P. spadiciformis*, *P. striata*, *Macrostachya noei*, *M. pendulata*, *Kallostachys scottii*, *Huttonia spicata*, *Weissistachys kentuckiensis*, *Cingularia typica*, *Calamocarpon insignis*, *Pendulostachya cinguliformis*
Elaterites	*Calamostachya americana*, *C. inversibractus*, *Palaeostachya andrewsii*, *P. decacnema*, *Macrostachya noei*, *M. pendulata*, *Weissistachys kentuckiensis*, *Calamocarpon insignis*, and *Pendulostachys cinguliformis*
	Sphenophyllaleans
Calamospora	*Bowmanites moorei*, *B. myriophyllus*, *B. nindelii*, *B. priveticensis*, *B. stimulosus*, *B. verticillatus*, *B.* sp. (*herein*), *B. tenerrimum*, *Sphenophyllum aguensis*, *S. beinertii*, *S. waldenburgense*, *Cheirostrobus pettycurensis*, *Pothocites grantonii*, and *P. pettycurensis*
	Noeggerathialeans
Calamospora (micro- and megaspores)	*Discinites bohemicus*, *D.* sp. (cf. *bohemicus*), *D. delectus*, *D. hanchengensis*, *D. hlizae*, *D. major*, *D. nemejcii*, *D.* cf. *raconicensis*, *D. vicinalis*, *Lacoea seriata*, *Paratingia wudensis*, *Tingia unita*, *T* sp., *Tingiostachya tetralocularis*, and *T.* sp., *Noeggerathiaestrobus* (only megaspores)
Verrucosisporites	*Noeggerathiaestrobus bohemicus*

Calamospora was produced by several parent plant species of different affinity [7], but its main producers were Pennsylvanian calamiteans. The first record of the in situ spores was interpreted as *Calamospora* (*Calamospora atava* (Naumova) McGregor and *C. pannucea* Richardson) or *Retusotriletes* type) and is known as the zosterophyll species *Sawdonia acanthotheca* Gensel et al. [41]. Note that *Retusotriletes* is curvaturate and, therefore, distinct from *Calamospora* and that most of the zosterophylls possess *Retusotriletes*-type spores. *Calamospora*-type spores, which are reported in some zosterophylls, such as *S. acanthotheca*, may be immature. From the Paleozoic plants of possible lycophyte affinity, we know only one record of an in situ *Calamospora*, Mississippian genus *Eleutherophyllum* Stur [42].

Other non-calamitean *Calamospora* records [7] include the Triassic *Bustia ludowici* Grauvogel-Stamm, *Echinostachys cylindrica* Schimper and Mougeot, *E. oblongus* Brongniart (*Calamospora keuperiana* and *C. mesozoica* types), *E. verticillata* Grauvogel-Stamm (*Calamospora tener* type), *Equisetostachys nathorstii* Halle, *E. suecius* (Nathorst) Halle (*Calamospora mesozoica* type), and Devonian *Protobarinophyton obrutschevii* Ananiev (*Calamospora atava* type.

8. Conclusions

A comparison of hundreds of palynological slides with in situ populations, isolated from sixty-one specimens of Pennsylvanian calamitean cones belonging to sixteen species of four genera, allows the definition of morphological variations of in situ *Calamospora*. The classification of dispersed *Calamospora* is purely arbitrary because all morphological criteria, including the diameter, thickness of exine, number, shape, position, and size of the secondary folds of the exine, as well as the length of rays of the trilete marks, are variable within one in situ *Calamospora* population. Some features are variable, and others are related to different ontogenetic stages.

The calamitean cones matured gradually from the base to the apex, as demonstrated by the different diameters of spores and a more frequent occurrence of contact areas on the spores isolated from basal and apical sporangia. The calamitean cones were bisporangiate, and an arbitrary size criterion for the division of micro- and megaspores (200 µm) may not be biologically meaningful. Some cones yielded spores with a bimodal size distribution, so even though the larger spores were less than 200 µm, they probably represent megaspores because the size in these populations exhibits a bimodal curve.

There is no evidence that *Elaterites* and *Vestispora* are ontogenetic stages of *Calamospora*, as demonstrated by the morphological and stratigraphical differences. However, it is evident that some Paleozoic calamitaleans produced not only *Calamospora* but also the *Elaterites* spores.

The majority of Paleozoic *Calamospora* producers were calamitean cones, mainly the genera *Calamostachys*, *Palaeostachya*, and *Macrostachya*. Sphenophyllaleans were a minor source of *Calamospora*. Some noeggerathialeans also produced *Calamospora*.

Author Contributions: Conceptualization, J.B.; Methodology, J.B.; Formal analysis, J.B.; Investigation, J.B. and J.V.F.; Data curation, J.V.F.; Project administration, J.V.F. All authors have read and agreed to the published version of the manuscript.

Funding: This research was supported by the Grant Agency of the Czech Republic (19-06728S) and the research plan of the Institute of Geology of the Czech Academy of Sciences (RVO67985831).

Institutional Review Board Statement: No applicable.

Informed Consent Statement: No applicable.

Data Availability Statement: Material including plant specimens is stored in the National Museum, Prague, Czech Republic and palynological slides are in the Institute of Geology, Academy of Sciences of the Czech Republic, Prague, Czech Republic.

Acknowledgments: We are much obliged to C. Cleal, University of Bristol, UK, for the linguistic revision and helpful remarks.

Conflicts of Interest: The authors declare no conflicts of interest.

References

1. Bonamo, P.M.; Banks, H.P. *Calamophyton* in the Devonian of New York state. *Am. J. Bot.* **1965**, *53*, 778–791. [CrossRef]
2. Feng, S.N.; Ma, J. Studies on the genus *Hamatophyton*. *Acta Bot. Sin.* **1991**, *33*, 140–146, (In Chinese with English Abstract).
3. Huang, P.; Liu, L.; Deng, Z.; Basinger, J.F.; Xue, J. *Xihuphyllum*, a novel sphenopsid plant with large laminate leaves from the Upper Devonian of South China. *Palaeogeogr. Palaeoclimatol. Palaeoecol.* **2017**, *466*, 7–20. [CrossRef]
4. Schweitzer, H.J. Die Oberdevon-flora der Bäreninsel. I. *Pseudobornia ursina. Nathorst. Palaeontographica* **1967**, *B123*, 116–137.
5. Taylor, T.N.; Taylor, E.L.; Krings, M. *Paleobotany: The Biology and Evolution of Fossil Plants*, 2nd ed.; Academic Press: Cambridge, MA, USA, 2009; p. 1231.
6. Pryer, C.M.; Schneider, H.; Cranfill, R.; Wolf, P.G.; Hunt, J.S.; Sipes, S.D. Horsetails and ferns are a monophyletic group and the closest living relatives to seed plants. *Nature* **2001**, *409*, 618–622. [CrossRef]
7. Nishiyama, T. Evolutionary developmental biology of nonflowering land plants. *Int. J. Plant Sci.* **2007**, *168*, 37–47.
8. Balme, B.A. Fossil in situ spores and pollen grains: An annotated catalogue. *Rev. Palaeobot. Palynol.* **1995**, *87*, 81–323.
9. Schopf, J.M.; Wilson, L.R.; Bentall, R. An annotated synopsis of Paleozoic fossil spores and the definition of generic groups. *Ill. St. Geol. Surv. Rep. Investig.* **1944**, *91*, 1–82.
10. Trivedi, G.K. Palynofloral diversity in the Kopili Formation (Late Eocene) from north-east India. *Acta Mus. Nat. Prag. Ser. B Hist. Nat.* **2009**, *65*, 9–24.
11. Good, C.W. Taxonomic and stratigraphic significance of the dispersed spore genus *Calamospora*. In *Geobot*; Romans, R.C., Ed.; Plenum: New York, NY, USA, 1977; pp. 43–64.
12. Bek, J. Spore Populations of Some Plants of Groups Lycophyta, Sphenophyta, Pteridophyta and Progymnospermophyta from Carboniferous Limnic Basins of the Czech Republic. Ph.D. Thesis, Geological Institute of the Academy of Sciences of the Czech Republic, Prague, Czech Republic, 1998. (In Czech).
13. Good, C.W.; Taylor, T.N. The morphology and systematic position of Calamitean elater-bearing spores. *Geosci. Man.* **1975**, *11*, 133–139. [CrossRef]
14. Němejc, F. Taxonomical studies on the fructifications of the Calamitaceae collected in the coal districts of Central Bohemia. *Sbor. Nár. Mus. B* **1953**, *9*, 1–62.
15. Boureau, É. *Traité de Paléobotanique. Tome III. Sphenophyta, Noeggerathiophyta*; Masson et Cie: Paris, France, 1964; pp. 1–545.
16. Hartung, W. Die Sporenverhaltnisse der *Calamariaceen. Arb. Inst. Paläobot. Petrogr. Brenn.* **1933**, *3*, 95–149.
17. Pešek, J.; Holub, V.; Jaroš, J.; Malý, L.; Martínek, K.; Prouza, V.; Spudil, J.; Tásler, R. *Geologie a Ložiska Svrchnopaleozoických Limnických Pánví ČESKÉ Republiky*; Český Geologický Ústav: Praha, Czech Republic, 2001.
18. Opluštil, S.; Schmitz, M.; Cleal, C.J.; Martínek, K. A review of the Middle-Late Pennsylvanian west European regional substages and floral biozones, and their correlation to the Global Time Scale based on new U-Pb ages. *Earth Sci. Rev.* **2016**, *154*, 301–335. [CrossRef]
19. Tásler, R.; Čadková, Z.; Dvořák, J.; Fediuk, F.; Chaloupský, J.; Jetel, J.; Kaiserová-Kalibová, M.; Prouza, V.; Skřivánková-Hrdličková, D.; Středa, J.; et al. *Geology of the Bohemian Part of the Intra-Sudetic Basin*; Academia: Praha, Czech Republic, 1979; p. 296. (In Czech with English summary)
20. Opluštil, S.; Šimůnek, Z.; Pšenička, J.; Bek, J.; Libertín, M. A 25 million year macrofloral record (Carboniferous–Permian) in the Czech part of the Intra-Sudetic Basin; biostratigraphy, plant diversity and vegetation patterns. *Rev. Palaeobot. Palynol.* **2017**, *244*, 241–307. [CrossRef]
21. Punt, W.; Hoen, P.P.; Blackmore, S.; Nilsson, S.; LeThomas, A. Glossary of pollen and spore terminology. *Rev. Palaeobot. Palynol.* **2007**, *143*, 1–81. [CrossRef]
22. Potonié, R.; Kremp, G. Die Gattungen der paläozoischen *Sporae dispersae* und ihre Stratigraphie. *Geo Jahr* **1954**, *69*, 111–193.
23. Potonié, R.; Kremp, G. Die *Sporae dispersae* des Ruhrkarbons, ihre Morphographie und Stratigraphie mit Ausblicken auf Arten anderer Gebiete und Zeitabschnitte. Teil I. *Palaeontogr. B* **1955**, *98*, 65–121.
24. Smith, A.H.; Butterworth, M.A. Miospores in the coal seams of the Carboniferous of Great Britain. *Spec. Pap. Palaeontol.* **1967**, *1*, 1–324.
25. Good, C.W. Pennsylvanian-age calamitean cones, elater-bearing spores, and associated vegetative organs. *Palaeontogr. B* **1975**, *153*, 28–99.
26. Barthel, M. Die Rotliegendflora Sachsens. *Abh. Staat. Mus. Miner. Geol. Dresden.* **1976**, *24*, 1–190.
27. Serret, L.; Brousmiche, C. Quelques fructifications de Calamitaceae (Arthrophytes recoltées dans le bassin houiller sarro-lorrain (organisation—Sporae in situ). *Palaeontogr. B* **1987**, *203*, 135–179.
28. Coquel, R.; Brousmiche-Delcambre, C. Comparaisons spores in situ-spores dispersées chez quelques Equisétales, Marattiales et Lépidodendrales du Carbonifere supérieur: Considérations sur l'utilisation de morphoespeces en Paléopalynologie. *Rev. Paléobiol.* **1996**, *15*, 121–154.
29. Libertín, M.; Bek, J. Proposal of the new classification of Palaeozoic sphenophyllalean cones. In Proceedings of the Abstracts of 7th European Palaeobotany Palynology Conference, Prague, Czech Republic, 6–11 September 2006; p. 82.
30. Libertín, M.; Bek, J. Bolsovian *Calamostachys incrassata* (Němejc) emend. and its spores from the Kladno-Rakovnik Basin of the Czech Republic. *Bull. Geosci.* **2006**, *81*, 207–213. [CrossRef]
31. Wilson, L.R. Elater-bearing spores from the Pennsylvanian strata of Iowa. *Am. Midl. Nat.* **1943**, *30*, 518–523. [CrossRef]
32. Wilson, L.R. Palynological assemblage resemblance in the Croweburg coal of Oklahoma. *Geol. Notes* **1964**, *24*, 138–143.

33. Peppers, R.A. Correlation and palynology of coals in the Carbondale and Spoon Formations (Pennsylvanian) of the northeastern part of the Illinois Basin. *Ill. Geol. Surv. Bull.* **1970**, *93*, 1–173.
34. Ravn, R.L. *An Introduction to the Stratigraphic Palynology of the Cherokee Group (Pennsylvanian) Coals of Iowa*; Iowa Geological Survey: Iowa City, IA, USA, 1979; Volume 6, p. 117.
35. Good, C.W.; Taylor, T.N. The establishment of *Elaterites triferens* spores in *Calamocarpon insignis* microsporangia. *Trans. Am. Microsc. Soc.* **1974**, *93*, 148–151. [CrossRef]
36. Ravn, R.L. Paleobotanical relationship and stratigraphic importance of the Carboniferous miospore genus *Vestispora* and questionably allied genera. *J. Paleontol.* **1983**, *57*, 568–580.
37. Bek, J. Paleozoic in situ spores and pollen. Sphenopsida. *Palaeontogr. B* **2021**, *301*, 141–201. [CrossRef]
38. Bek, J.; Dimitrova, T. Taxonomy and stratigraphic importance of the Carboniferous miospore genus *Vestispora*. *Rev. Palaeobot. Palynol.* **2013**, *174*, 122–135. [CrossRef]
39. Kurrmann, M.H.; Taylor, T.N. Comparative ultrastructure of the sphenophyte spores *Elaterites* and *Equisetum*. *Grana* **1984**, *23*, 109–116. [CrossRef]
40. Lugardon, B.; Brousmiche-Delcambre, C. Exospore ultrastructure in Carboniferous sphenopsids. In *Ultrastructure of Fossil Spores and Pollen*; Kurmann, M.H., Doyle, J.A., Eds.; The Royal Botanic Gardens: Kew, London, 1994; pp. 53–66.
41. Gensel, P.G.; Andrews, H.N.; Forbes, W.H. A new species of *Sawdonia* with notes on the origin of microphylls and lateral sporangia. *Bot. Gaz.* **1975**, *136*, 50–62. [CrossRef]
42. Remy, R.; Remy, W. *Eleutherophyllum drepanophyciforme* n. sp. aus Namur A von Niederschlesien. *Sencken. Lethaia* **1960**, *41*, 89–100.

life

MDPI

Article

Multiple Non-Destructive Approaches to Analysis of the Early Silurian Chain Coral *Halysites* from South China

Xinyi Ren [1], Yazhou Hu [1,*], Peiyu Liu [2], Yue Liang [1], Feiyang Chen [3], Hao Qiu [4], Luke C. Strotz [1,5], Kun Liang [6] and Zhifei Zhang [1]

1 State Key Laboratory of Continental Dynamics, Shaanxi Key Laboratory of Early Life and Environments, Department of Geology, Northwest University, Xi'an 710069, China; 201810275@stumail.nwu.edu.cn (X.R.); yue.liang@nwu.edu.cn (Y.L.); lukestrotz@nwu.edu.cn (L.C.S.); elizf@nwu.edu.cn (Z.Z.)

2 Key Laboratory of Deep Petroleum Intelligent Exploration and Development, Institute of Geology and Geophysics, Chinese Academy of Sciences, Beijing 100029, China; liupeiyu@mail.iggcas.ac.cn

3 School of Resources and Geosciences, China University of Mining and Technology, Xuzhou 221116, China; feiyang.chen@cumt.edu.cn

4 College of Architecture and Geomatics Engineering, Shanxi Datong University, Datong 037003, China; qiuhao@sxdtdx.edu.cn

5 Biodiversity Institute and Department of Ecology & Evolutionary Biology, University of Kansas, Lawrence, KS 66045, USA

6 State Key Laboratory of Palaeobiology and Petroleum Stratigraphy, Nanjing Institute of Geology and Palaeontology, Chinese Academy of Sciences, Nanjing 210008, China; kliang@nigpas.ac.cn

* Correspondence: eliyazhou@nwu.edu.cn

Abstract: Cnidarians are among the most important diploblastic organisms, elucidating many of the early stages of Metazoan evolution. However, Cnidarian fossils from Cambrian deposits have been rarely documented, mainly due to difficulties in identifying early Cnidarian representatives. *Halysites*, a tabulate coral from Silurian reef systems, serves as a crucial taxon for interpreting Cambrian cnidarians. Traditionally, the biological characteristics of *Halysites* have been analyzed using methods limited by pretreatment requirements (destructive testing) and the chamber size capacity of relevant analytical instruments. These constraints often lead to irreversible information loss and inadequate data extraction. This means that, to date, there has been no high-resolution three-dimensional mineralization analysis of *Halysites*. This study aims to introduce novel, non-destructive techniques to analyze the internal structure and chemical composition of *Halysites*. Furthermore, it seeks to elucidate the relationship between coral organisms and biomineralization in reef settings and to compare Silurian Tabulata with putative Cambrian cnidarians. Techniques such as micro-X-ray fluorescence spectrometry (micro-XRF), micro-X-ray computed tomography (micro-CT), and scanning electron microscopy (SEM) were employed in this research. With the help of high-resolution micro-CT scanning, we identify the growth pattern of *Halysites*, showing both lateral and vertical development. The lateral multiple-branching growth pattern of *Halysites* corals is first established herein. The flaggy corallite at the initial stage of branching is also observed. The micro-XRF mapping results reveal the occurrence of septa spines for *Halysites*, a trait previously thought rare or absent. Additionally, the ratio of coral volume to the surrounding rock was assessed, revealing that *Halysites* reefs were relatively sparse (volume ratio = ~30%). The cavities between *Halysites* likely provided more space for other organisms (e.g., rugose corals and bryozoans) when compared to other coral reef types. Additionally, we provide a comparative analysis of post-Cambrian colonial calcareous skeletons, offering insights into the structural features and growth patterns of early skeletal metazoans across the Ediacaran–Cambrian boundary.

Keywords: Tabulata corals; *Halysites*; micro-XRF; micro-CT

1. Introduction

Cnidarians are considered one of the most basal metazoan groups, representative of the earliest stages of Eumetazoan evolution. However, soft-bodied diploblastic fossils are infrequently found in pre-Cambrian and Cambrian deposits [1–4], meaning the earliest stages of cnidarian evolution have been difficult to ascertain. Corals, which represent significant reef-builders throughout the fossil record, secrete a mineralized exoskeleton similar to those produced by stromatoporoids and archaeocyathids, forming calcareous skeletons or spicules to reinforce their body structure [5–7]. One of the more common diploblastic skeletal clades found in the fossil record is the Tabulata, which were prevalent during the Paleozoic era, from the early Ordovician to the Permian, and are distributed across the globe [4,7–14].

Halysites is an important genus of Tabulata, having previously been used as a model taxon for interpreting the affinities of Cambrian metazoan taxa in carbonate deposits [1,15–17]. Characteristic of Tabulata, *Halysites* exhibit developed tabulae with weakly or slightly developed septa [17–19]. Previous researchers have examined the morphology, classification, and evolutionary significance of *Halysites* using thin sections and have documented its distribution and stratigraphic occurrences [16,20–24]. However, more recent research on *Halysites* has focused less on its biology, with a large portion of recent studies concentrating on its utility in oil and gas exploration [7–10,12,13,18,22,25]. Traditional methods of analyzing *Halysites* have relied on hand samples, polished slabs, and thin sections, techniques that are usually destructive [20,26,27], resulting in the loss of potential details and an inability to conduct multi-dimensional analyses.

To extract more information, advanced near-non-destructive technologies and methods, such as scanning electron microscopy (SEM), micro-X-ray fluorescence spectrometry (micro-XRF) and micro-X-ray computed tomography (micro-CT) were employed on well-preserved early Silurian examples of *Halysites* from the upper Xiangshuyuan Formation, Guizhou Province of South China (Figure 1), to investigate their 2D and 3D morphological characteristics. Additionally, this study utilizes Silurian Tabulata as a model to identify potential Cnidarians in Cambrian small shelly fossil faunas from comparable carbonate deposits.

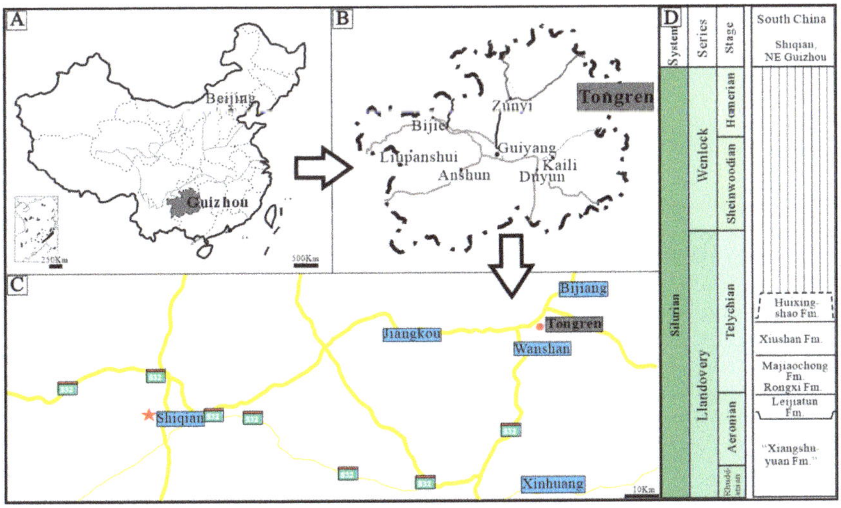

Figure 1. Location of the Shiqian section (red star) and its associated stratigraphic section. (**A**) the location of the study area in China (gray area); (**B**) the location of the study area in Guizhou Province (marked with a gray rectangle); (**C**) a route map of the sampling area (red-star marking the studied area) (modified from [28–30]); (**D**) stratigraphic column for the upper part of the Xiangshuyuan Formation in the early Silurian (modified from [17]).

2. Materials and Methods

More than 100 *Halysites catenularius* (Linnaeus, 1767) [31] specimens were collected from dark gray or grayish white limestones of the upper part of the Xiangshuyuan Formation, early middle Silurian, Shiqian County, Tongren, Guizhou of China (Figure 1). Four larger samples were selected for this research; two of them were selected for non-destructive instrumental analysis, and the other two were used to create thin sections. The first sample contains 172 coral individuals, and the second sample contains 179 coral individuals. All specimens are deposited in the collections of the Department of Geology of Northwest University, Xi'an, China. Fossils were examined under the binocular Zeiss Zoom Stereo microscope (Zeiss, Oberkochen, Germany) and were photographed using a stereo photographic Zeiss Smart Zoom 5 (Zeiss, Oberkochen, Germany). Three-dimensional non-destructive internal micro-structure was obtained using a Zeiss Xradia Versa 520 micro-CT (micro-X-ray computed tomography) (Zeiss, Oberkochen, Germany) [32,33], with a spatial resolution of 0.7 μm and 70 nm voxel imaging. The micro-CT data were reconstructed using the Dragonfly software 3.1. The surface structure of the corals was assessed using scanning electron microscopy (Quanta 400 FEG; FEI, Hillsbora, America). To obtain the chemical composition of *Halysites* specimens, micro-X-ray fluorescence spectrometry (micro-XRF) was used (Brucker, Karlsruhe, Germany), including mapping elemental analysis, which was used to compare the elemental differences between the fossil and surrounding rock, as well as the different parts of the fossils themselves [34–43]. The proportion of area occupied by *Halysites* was determined using micro-XRF images and using the ImageJ software 1.48 [31,44–46].

3. Results

3.1. Two-Dimensional Morphology and Mineralized Structure of Halysites

Halysites catenularius (Linnaeus, 1767) develops polygonal lacunae (Figure 2E), and rank junctions occur at tubules (Figure 2C). Additionally, corallites are subelliptical (Figure 2E–K). *Halysites* colonies are always arranged as chains composed of many individuals and are all linked by coenenchymal tubules (small tubes). Colonies often produce new individuals on both sides of the coral's long axis. The individual corallites are long and thin tubes, round or oval in shape, and the corallite walls (Figure 2A,B,F,J) consist of two layers: the outer wall, which is in contact with the external environment, and an inner wall, which surrounds the tubule chamber. The crystallization of the corallite walls compared to the coral cavity is notably different (Figure 2J). The fossilized parts of *Halysites* are gray-white, the matrix is dark gray, and *Halysites* individuals are connected by tubules, forming variously sized meshes or longitudinal and transverse tabula (Figure 2B).

The shape of the tubules is variable (quadrilateral or triangular, predominantly quadrilateral), and they are filled with calcite material similar to the calcite that fills the main chambers. Numerous irregular fractures are observed in the longitudinal section (Figure 2C,D). The tabulae are fully developed, straight, or slightly concave in the middle part of the colony. The fill between the tabulae also has numerous cracks, but no cracks develop on the tabula, and the cracks in the surrounding matrix do not cross them. The tubules are multilayered longitudinally, with the longitudinal single tube not possessing a uniform width and occasionally tapering abruptly upward. No septal spines were observed under SEM or polarizing microscopy (Figures 2 and 3).

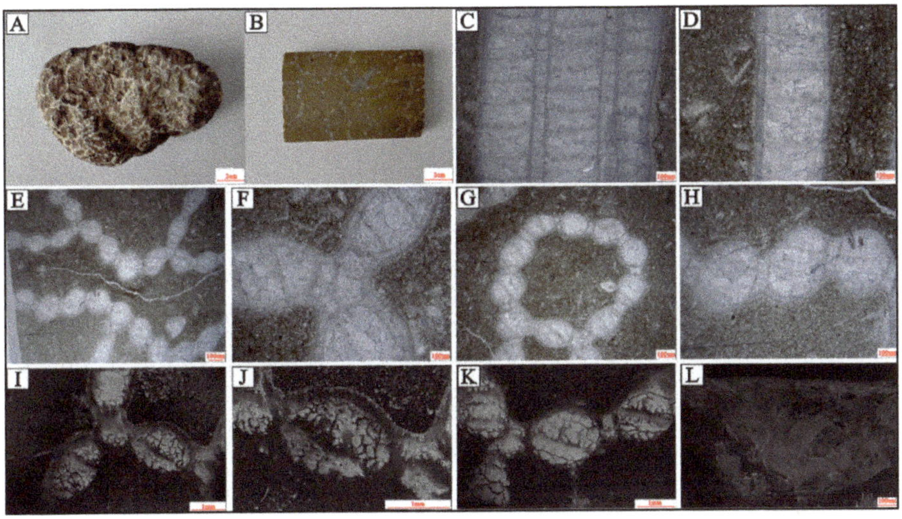

Figure 2. *Halysites* from the upper part of the Xiangshuyuan formation at Shiqian County, Guizhou, South China. The corallites are closely arranged, and corallite walls are relatively thick. (**A**,**B**) external view and photo of polished slab, (**C**,**D**) longitudinal sections under the polarized light, (**E**–**H**) transverse sections under the polarized light, and (**I**–**L**) transverse section images taken with SEM. *Halysites* corallites are connected in a chain-like pattern, with no fixed angle or number. Some form lacunae with up to 10 corallites, while others may extend further away. Tabulae are well-developed, flat, or slightly concave, and tubules share thick walls with corallites. The corallite's cavity fills have structures similar to the septa.

3.2. Three-Dimensional Morphology of Halysites

Halysites form irregular lateral ranks or lacunae, which are organized into multiple longitudinal columns and are closely connected laterally to create a fence-like structure. The examined *Halysites* fossils are well preserved, consisting of complete three-dimensional forms with no signs of fragmentation or fracture. Due to the similarity between coral skeletons and associated infilling material, *Halysites*' internal structures, such as connecting tubules, cannot be identified. The *Halysites* chain extends horizontally and connects to surrounding *Halysites*, exhibiting a horizontal layered growth pattern (Figures 3–7). Some specimens grow in a straight line, while others bend or grow in layers. At the top, they continue to grow in the vertical direction. Each group of monomers can connect to two or even four adjacent groups of *Halysites*. *Halysites* develop vertically in an upward direction and reproduce at certain heights. Specifically, the coral's tube splits around the middle of the coralla. Micro-CT shows that as they grow upwards, the corallum becomes larger compared to the base, and the area at the highest point is also wider, as indicated by the white arrow (Figure 4E). *Halysites* forms in chains but still possess the short, columnar, longitudinally stacked shape that is characteristic of the Tabulata. Consequently, the mesh morphology of *Halysites* coralla is diverse and flexible.

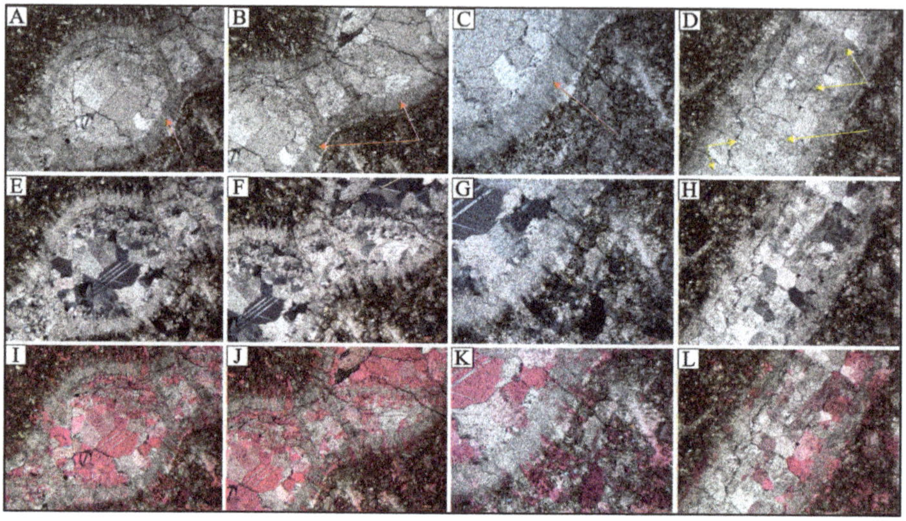

Figure 3. Photomicrographs of *Halysites* captured under polarized light (**A–D**) and polarized microscope images (**E–L**) using different test plates to distinguish the structure of each part. (**A**,**E**,**I**) represent individual corallites (red arrows) with tubules on two sides, and the matrix is brown, while the corallite wall is dark gray, and the fill consists of coarse-grained crystals. The tubules wall and corallites wall are composed of fine mineral particles, and their fills have a similar composition. (**B**,**F**,**J**) show corallites are connected by tubules, and tubules are rectangle-shaped. The tubule wall is thinner than the corallite wall. (**C**,**G**,**K**) show the boundary of corallites and the surrounding matrix. The fill consists of coarse calcite particles, the corallite wall is composed of very fine particles, and the matrix has a mixture of minerals and particle sizes. (**D**,**H**,**L**) show the composition and particle size for tabulas (yellow arrows), fills, and matrices. The tabulas consist of fine particles, similar to the corallite walls (red arrows), and corallite walls are thicker than the tabulas.

3.3. The Elemental Analysis of Halysites

Elemental distribution mapping was performed using micro-XRF to obtain semi-quantitative elemental distributions (Figure 5). Results of these analyses show that, although Ca is distributed in both fossil and matrix regions, the region occupied by the fossil has a stronger calcium signal compared to the matrix (Figure 5A,D,G). The Ca signal (represented in yellow false color) is markedly stronger in the coral structures, while Si and Fe signals (represented in blue and red false colors, respectively) were weak in the regions occupied by the coral and much stronger in the matrix (Figure 5A,C). Analysis of the mapping tests indicates that the septal spines are well preserved, and it can be seen that inside each coral individual, there is a calyx-like structure, but this feature is difficult to distinguish in hand specimens or under polarized light. Micro-XRF mapping of Fe and Ca highlights the position and distribution of the septal spines. Septal spines are very irregular in shape, and some are divided into two or three parts. The septal spines are short and thin, and some corallites do not have septal spines. The size and number of septal spines varies (4–12 septal spines). The minimum number is due to internal space splitting, resulting in 4–6 septal spines. In addition, there are also structures similar to septal spines in the connecting tubules that are sometimes connected to each other, with 2–3 possible radiating arrangements (Figure 5K,L). The length of the septal spines ranges from 0.094 mm to 0.450 mm.

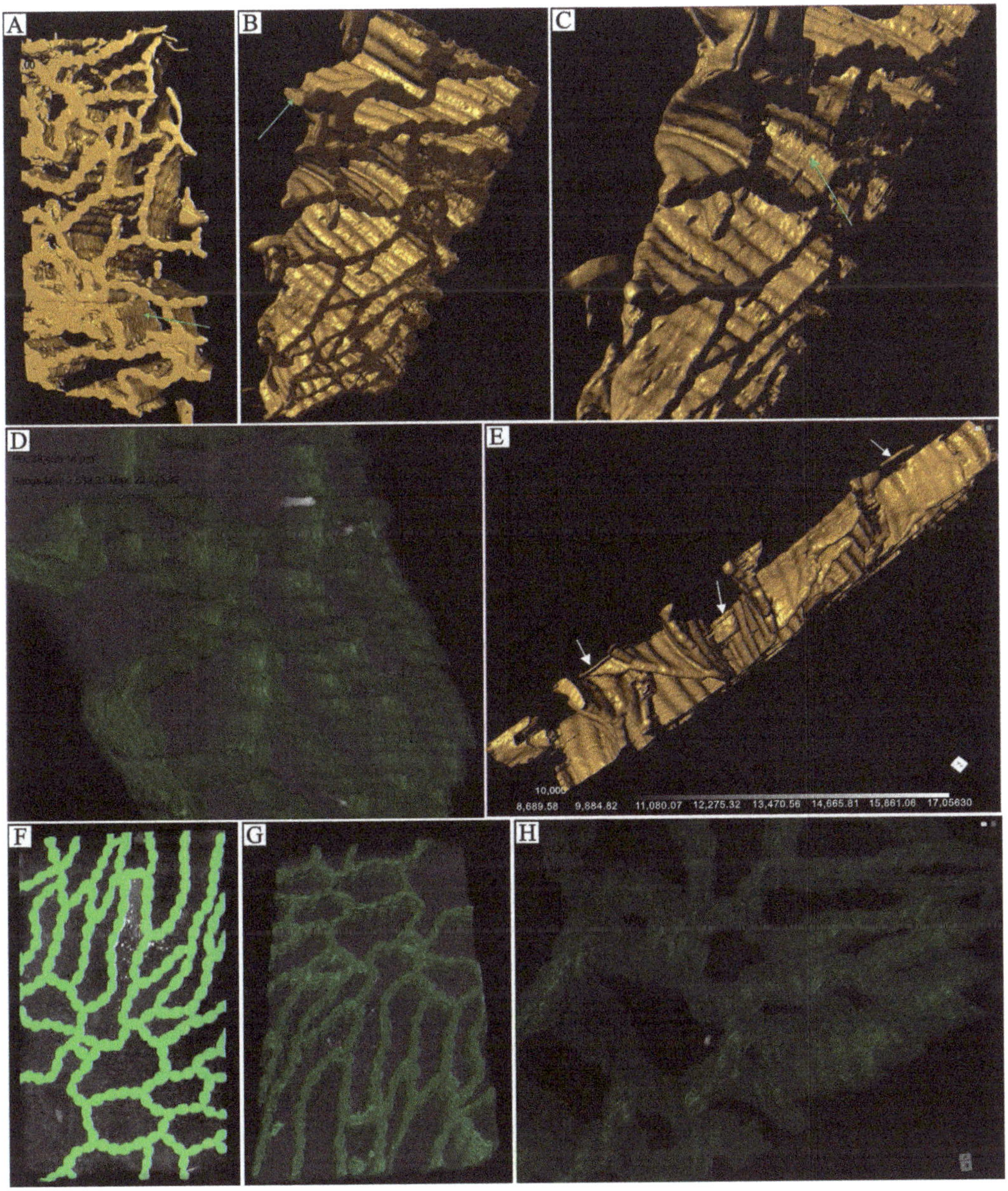

Figure 4. Three-dimensional reconstruction of *Halysites* based on micro-CT. The resolution of the first set of micro-CT images is 38.22 μm/pixel (images with yellow false color), and for the second set (green false color), the resolution is 41.13 μm/pixel. (**A–C**,**E**) display corals (yellow) extracted from the matrix; (**D**,**F–H**) depict corals (light gray) against a gray background; (**A–C**) illustrate the shape of the coral when connected with others, and tabula that are oriented in a longitudinal direction can be seen (green arrow); (**E**) newly forming branches at the top of the specimen (white arrow).

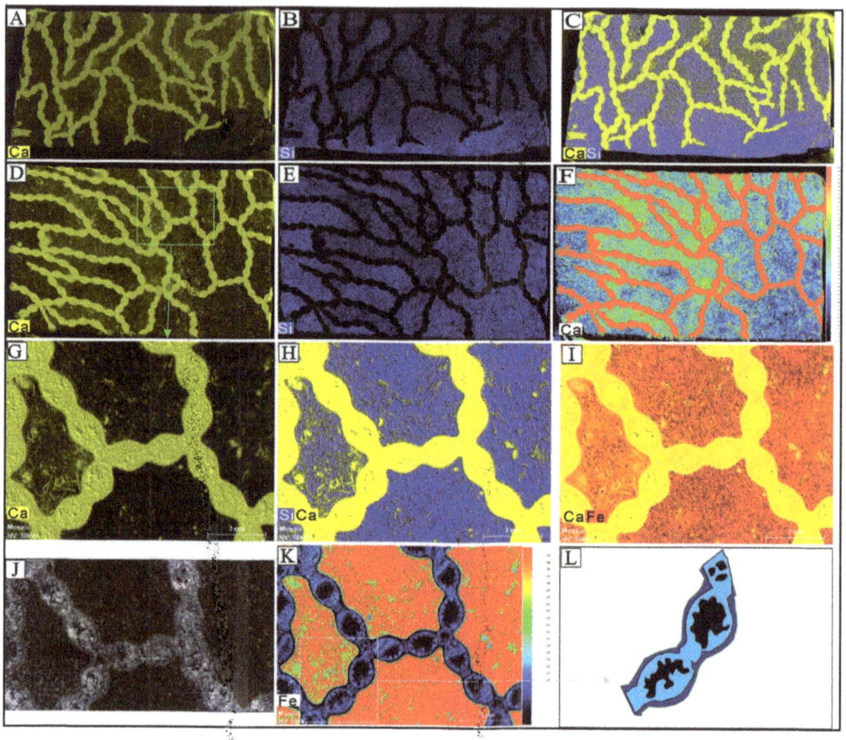

Figure 5. Elemental mapping of *Halysites* skeleton and surrounding sedimentary matrix using micro-XRF imaging (Ca—calcium, Fe—iron, Si—Silicon). (**A**,**D**,**G**) distribution maps of Ca, G located in the green rectangle of D; (**B**,**E**) distribution maps of Si; (**C**,**H**) coupled distribution maps of Si–Ca; (**J**) polished section of *Halysites*; (**I**) coupled distribution maps of Fe–Ca; (**F**) heatmap showing the distribution of Ca abundance; (**K**) heatmap showing the distribution of Fe abundance; (**L**) a schematic drawing of the coral calyx. The content of the light gray box in (**K**) is interpreted to represent the remains of coral polyps and septal spines. Purple represents the coral tube wall, blue represents the coral calyx, and black represents the coral cavities.

3.4. *Halysites* Reef Volume Calculation

For volume calculations, more than 100 samples from similar preservation conditions were collected, and two larger slabs were selected, both containing multiple individuals connected in chains, with the chosen slabs best fully representing the preservation and morphology of *Halysites*. ImageJ software was used to determine the proportion of fossils to the total rock area. The connecting tubules and coralla were distinguished in the software by coloring them black, with the surrounding rock represented by white. In the fossil areas, some white debris was mixed in, representing secondary sediment infilling. Similarly, black fragments were found in the surrounding rock, indicating that fossils had weathered or become fragmented and mixed with the rock. The software calculated the relative area occupied by fossil material on the first block to be 27.003%. The relative area of the second block is calculated to be 30.075% (Figure 6).

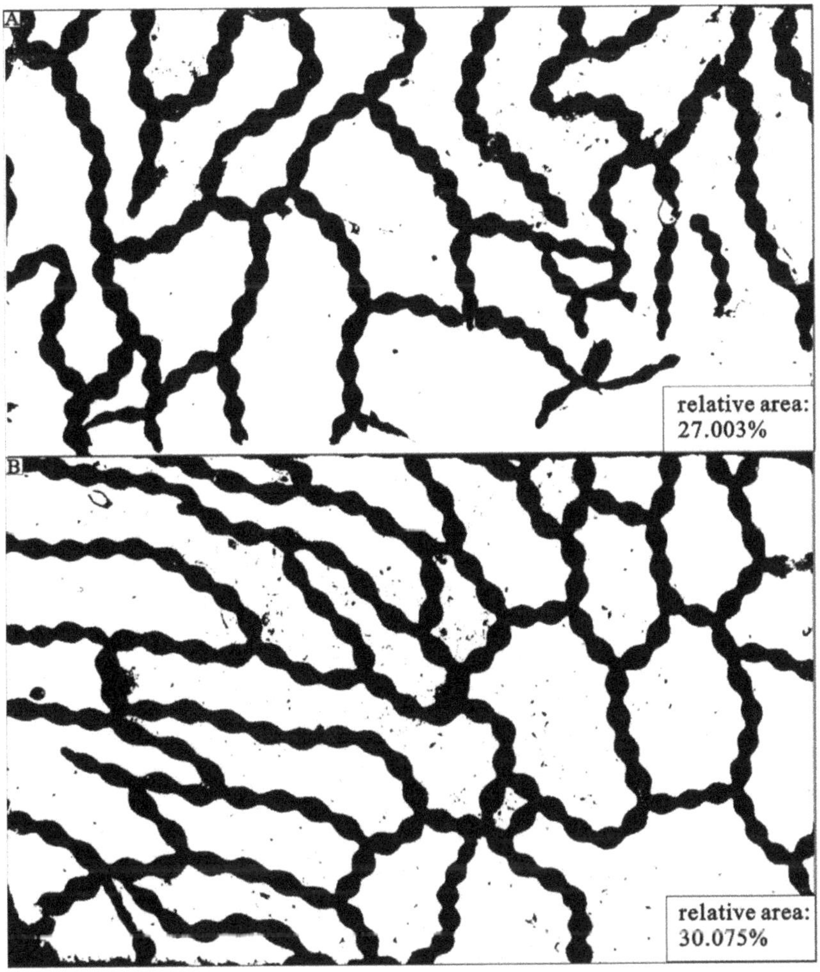

Figure 6. The relative area of *Halysites* versus the surrounding matrix. The black chains represent the coral skeleton, and the white represents the matrix component. In the matrix, there are some scattered broken coral fragments. (**A**) The area occupied by *Halysites* takes up 27.003%, and in (**B**) 30.075%.

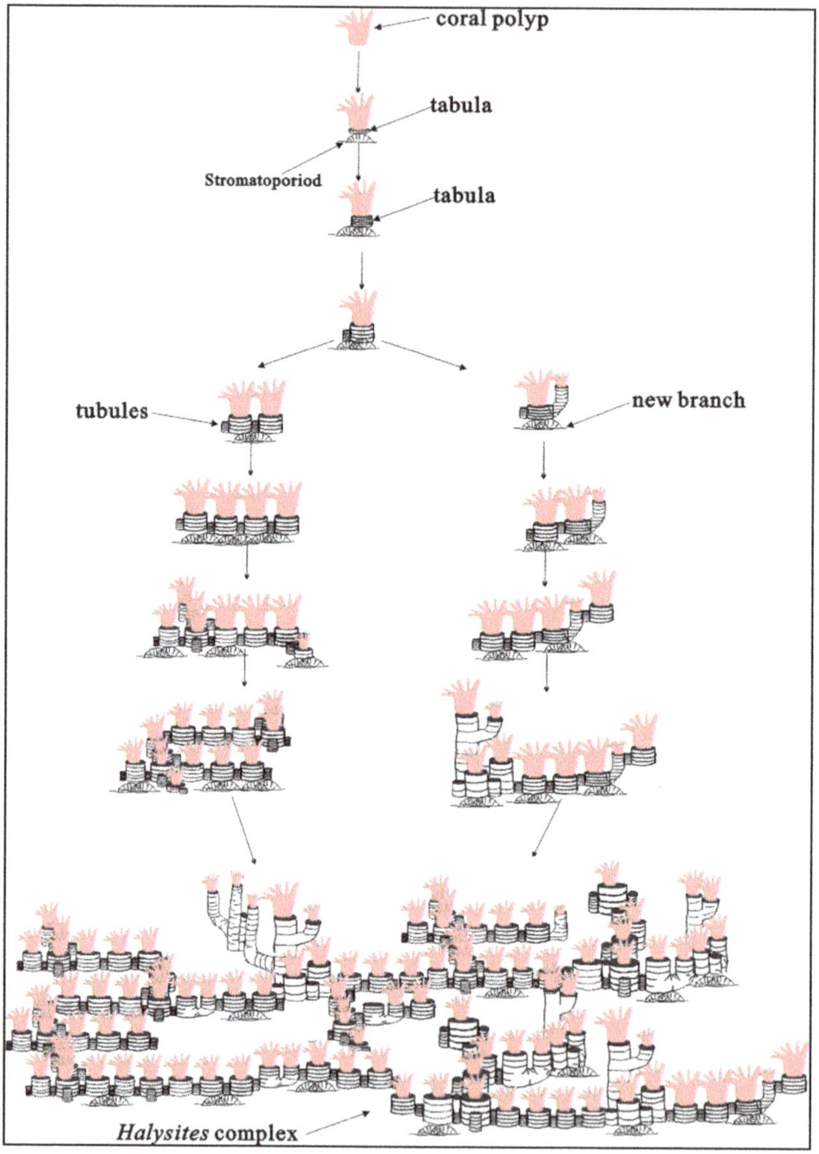

Figure 7. A reconstruction to illustrate the *Halysites* development process [47]. At the beginning, the coral larvae mature and begin to secrete calcite to produce the corallite exoskeleton, forming the first tabula of the *Halysites* colony. Coral polyps continuously produce layered horizontal tabulae, which ultimately lift polyps away from the original substrate. Meanwhile, with the help of tubules, the newly formed ranks branch in multiple directions, resulting in the complex structure typical of *Halysites* reefs.

4. Discussion

4.1. New Insights into 2D and 3D Morphological Reconstruction of Halysites

Halysitid corals first appeared in the early late Ordovician and disappeared in the end of the Silurian [31,48]. Although the strange chain-shaped halysitid corals are abundant [31,33,49], the 2D and 3D morphological reconstructions have not been well established. The development of the growth pattern of halysitid colonies has been the focus of previous studies, which were solely based on thin sections or polished slabs, resulting in many insights into halysitid corals' skeleton development [50]. No details about the development of the whole halysitids have been established. Although the CT scanning method has been used in halysitids studies, the resolution is low (with 125 μm/pixel in [51]), making it hard to make precise reconstructions of the whole halysitids coral. Meanwhile, the CT scanning by [51] focuses on the growth pattern but without only vertical development information, leaving no details on how halysitids coral growth in horizontally. The non-destructive method in 2D and 3D views (micro-CT and micro-XRF) by hand samples offers several advantages than solely based on thin sections or polished slabs (e.g., [31,51–53]). The high resolution herein (with 38.22 μm/pixel and 41.13 μm/pixel) reveals the high contrast between coral skeletons (coral cavity fills) and the surrounding matrix, providing an excellent view of the growth pattern of *Halysites* that is better than previous research. As mentioned before, branching growth patterns are always expected near the lateral edge of the whole chain in corals (Figure 4E). The single colony would be branched into two or three colonies, with the branching angle ranging from 60° to 120°. Meanwhile, there are also some branches occurring with only one colony, possibly representing the dying colony (Figure 6B). Considering the complete morphology revealed by micro-CT scanning, we proposed that branching exists in the coral colony skeleton structure and that branching is a growth pattern for reef expansion. Based on the high-resolution micro-CT scanning results, we made a reconstruction to illustrate the *Halysites* development process (Figure 7). In the beginning, the corallite will inhabit a hard support (e.g., shells or skeletons of other organisms) [5], and then the pre-existing rank will be elongated by the insertion of new coral zooids (Figure 7). After the vertical increase of corallite, followed by a lateral increase by branching the new ranks, it will be linked with old coral zooids by intercorallite dissepimental tissue [53]. With multiple laterally branching methods, a polygonal-shaped *Halysites* coral reef is formed (Figures 2B and 4–6).

Although micro-CT reveals coral structures in a high resolution. The contrast between the coral wall or coral original skeleton and the sparry calcite fills is low, resulting in no further detailed analysis by using micro-CT. Micro-XRF offers non-destructive elemental testing and has recently gained popularity in sedimentology, paleontology, and geology [34,35,37,38,54,55]. The fossilized corals studied herein are about 2 mm in diameter of the corallite and under the microscopic XRF field of view, showing high resolution of *Halysites* details (Figures 4 and 5). Micro-XRF analysis indicates the distribution of silica, calcium, and iron, revealing a five-part structure in chain corals (Figure 5K). Specifically, the iron element mapping results of micro-XRF show some irregular tiny structures in the coral zooids that could be the septal spines of *Halysites*. Septal spines in *Halysites* do not always exist [16,19,28,56], and this is particularly true for the observation from polished slabs and thin sections observed by microscope (Figures 2 and 3). Due to the different structures and functions, there must be subtle differences in the composition of the outer epithelial when secreting coral walls and septal spines. However, under the micro-XRF scanning, septal spines are intensely occurring in most *Halysites* coral zooids (Figure 5K,L), implying that there are limitations in traditional microscopic observation technology and micro-XRF could provide high-resolution internal structure analysis of *Halysites* coral zooids. This might be due to the ability of organic matter to adsorb iron elements in sediments and preserve them in the form of complexes that the abnormal iron signal is due to the residual organic carbon in the fossil burial process, which enriches the iron element in the fluid and forms hematite [57]. In other words, the iron signal shows the organic soft-bodied original position and the septal spines might be produced through coral's secretion and mineralization. Therefore, the internal skeleton structure could be revealed by different elemental comparisons in micro-XRF.

4.2. Implications for *Halysites* Reef Formation

Coral reefs have been regarded as one of the most important examples of ecosystem engineering, creating multiple niches for other organisms [17,19]. Coral reef formation in the Paleozoic is usually primarily a result of biomineralization by the coral themselves, although sometimes aided by other organisms such as stromatoporoids, pelmatozoans, and bacteria [58]. The initial growth of coral reefs always requires the presence of a hard substrate (e.g., shells, skeletons, or lithified rocks [44]). *Halysites* is no exception to this pattern, and the formation of *Halysites* reefs starts with a single corallite. As the colony expands, both vertically and laterally, a complex *Halysites* reef is formed. In our study area, the *Halysites* themselves only take up ~30% of the total available volume, with the cavity between *Halysites* chains filled with soft matrix (Figure 6). However, those cavities are not always filled with soft matrix in other study areas. A recent study of Silurian *Halysites* from the Lower Visby Formation, Gotland, Sweden, reveals that the halysitid colonies there have been used as a hard substrate for rugose corals to stabilize their position in what is a soft sediment environment [26]. The rugose corals also occupy the cavities between halysitid ranks. Compared with tabulate coral reefs constructed by other taxa, *Halysites* reefs seemingly provide relatively more space for other organisms to inhabit.

4.3. Implications for Small Shelly Fossils Assessment

Tabulate corals appeared in the middle Ordovician and diversified in the Silurian and Carboniferous periods. Cnidarians have been repeatedly reported from early Cambrian deposits, but no diploblast examples, e.g., tabulata, have been found [1–3,49,56]. There are many tubular fossils from Cambrian deposits that remain unclassified, and there may be traces of Tabulata, but evidence is lacking in mudstone or shale [49]. Consequently, it is advantageous to seek out three-dimensional preserved Eumetazoa in carbonates, where such fossils may be preserved in great detail and obtained through acid dissolution. After etching the carbonates, the phosphatized steinkerns are left, resulting in isolated elements of a single taxon that may be interpreted as different species. For example, cylindrical microtubules of likely corals were interpreted as stems and carpal branches of crinoids [59]. Similar microtubes have also just been classified as tubular fossils [60]. More recent work has revealed these tubes should be considered basal cnidaria and were similar to younger coral taxa [61–64]. These tubular fossils have similar structures to those observed for *Halysites* in our study (Figure 3), including a tectum, inner–outer tectoria, and a core filled with different minerals [65–67]. This suggests the possibility that some of these tubular fossils may be the unrecognized tabulata.

5. Conclusions

Multiple 2D and 3D nondestructive methods were performed on specimens of *Halysites* for the first time, allowing detailed analysis of *Halysites* morphology with no damage to the actual specimens. With the help of high-resolution micro-CT scanning, we identify a pattern of growth for *Halysites*, with both lateral and vertical development and evidence of multiple branching. The layered corallite at the initial stage of branching is also observed. Elemental distribution analysis of the *Halysites* skeleton, coral domatia, and surrounding matrix allows for clear differentiation of the coral morphology and ultrastructure. Results of micro-XRF mapping reveal the occurrence of large concentrations of septal spines in *Halysites* corallites that were previously thought rare or absent. Additionally, based upon an assessment of the volume occupied by *Halysites* colonies, it seems *Halysites* occupied only approximately 1/3 of the available space, with the cavities between *Halysites* chains providing space for other organisms (e.g., rugose coral and bryozoans). The reconstruction of *Halysites* herein may provide valuable reference data for seeking basal or stem-group cnidarians in Cambrian deposits.

Author Contributions: Conceptualization, Z.Z.; methodology, H.Q. and Z.Z.; formal analysis, X.R. and Z.Z.; investigation, X.R. and Z.Z.; resources, Z.Z.; data curation, X.R., H.Q. and K.L.; writing—original draft preparation, X.R.; writing—review and editing, X.R., Y.L., P.L., Y.H., F.C., L.C.S., K.L. and Z.Z.; supervision, Z.Z.; project administration, Z.Z.; funding acquisition, Z.Z. All authors have read and agreed to the published version of the manuscript.

Funding: Current work was supported by the National Key Research and Development Program of China [Grant Number 2023YFF0803601], the National Natural Science Foundation of China [Grant Number 42302010], the Department of Science and Technology of Shaanxi Province [2022TD-11].

Institutional Review Board Statement: Not applicable.

Informed Consent Statement: Not applicable.

Data Availability Statement: Data related to this study can be found in Zenodo (10.5281/zenodo.12804814).

Acknowledgments: We warmly thank Liu Hongfu, Liang Kun, Zhai Juanpin, Zhang Qian, Sun Jie, Chen Shuangshuang, Ting Li, Zhang Jing, Zhang Dongdong, Ran Yazhou, Li Yining, and Feng Danting for fossil preparation and technical assistance in the Shaanxi Key Laboratory of Early Life and Environments, Northwest University of Xi'an. Useful suggestions on data analysis were given by Luo Mei, Zhang Caibin, Wang Jiayue, Liu Fan, Fu Rao, Song Baopeng, Zhao Jingliang, Fang Ruisen, Xingchen Wu and Wang Zhao in Xi'an and Beijing. Special thanks to the editors and reviewers for their wonderful comments that greatly benefited the paper.

Conflicts of Interest: The authors declare no conflicts of interest.

References

1. Han, J.; Guo, J.F.; Ou, Q. Evolutionary framework of early Cambrian cnidarians from south China. *Earth Sci. Front.* **2020**, *27*, 67–78. [CrossRef]
2. Liao, W.H. Advance in study of the taxonomy of Cnidaria Palaeozoic corals. *Acta Palaeontol. Sin.* **2002**, *41*, 464–468. [CrossRef]
3. Shu, D.G.; Conway, M. New diploblasts from Chengjiang fossil lagerstätte. *Earth Sci. Front.* **2006**, *13*, 227–233.
4. Vinn, O. Phosphatic Biomineralization in Scyphozoa (Cnidaria): A Review. *Minerals* **2022**, *12*, 1316. [CrossRef]
5. David, J.B. *Paleoecology-Past, Present and Future*; Wiley-Blackwell: Hoboken, NJ, USA, 2015.
6. Ricci, F.; Rossetto Marcelino, V.; Blackall, L.L.; Kühl, M.; Medina, M.; Verbruggen, H. Beneath the surface: Community assembly and functions of the coral skeleton microbiome. *Microbiome* **2019**, *7*, 159. [CrossRef] [PubMed]
7. Michael, J.B.; David, A.T.H. *Introduction to Paleobiology and the Fossil Record*; Wiley-Blackwell: Hoboken, NJ, USA, 2009.
8. Bhargava, O.N.; Bassi, U.K. Silurian reefal buildups: Spiti-Kinnaur, Himachal Himalaya, India. *Facies* **1986**, *15*, 35–51. [CrossRef]
9. Gong, S.Y.; Li, Z.M.; Cai, S.W. Origin, Emigration and Paleogeography of Ordovician—Silurian Tabulata in Yangtze Region. *Earth Sci.-J. China Univ. Geosci.* **1997**, *22*, 129–134. (In Chinese, English Summary)
10. Jin, C.T. Sequence of Silurian bed coral in Huanggexi, Daguan, Yunnan. *Acta Palaeontol. Sin.* **1984**, *23*, 4–140. (In Chinese)
11. Wang, B.Y. Paleontological Characteristics of Tabulata and Heliolitoidea of Silurian in northern China and adjacent area. *Xinjiang Geol.* **1998**, *2*, 14. (In Chinese, English Summary)
12. Yu, C.M. *Coral Fossils in China: Fossils of Various Phyla in China*; Science Press: Beijing, China, 1963. (In Chinese)
13. Zhou, X.Y. A preliminary study on the biostratigraphy of Silurian Tabulata in Guizhou province. *Acta Geol. Sin.* **1983**, *4*, 34–44. (In Chinese, English Summary)
14. Zapalski, M.K. Parasites in Emsian-Eifelian Favosites (Anthozoa, Tabulata) from the Holy Cross Mountains (Poland): Changes of distribution within colony. *Geol. Soc. Lond. Spec. Publ.* **2009**, *314*, 125–129. [CrossRef]
15. Franc¸oise, D. Lower Cambrian archaeocyathan bioconstructions. *Gen. Palaeontol. (Palaeoecol.)* **2007**, *6*, 5–19. [CrossRef]
16. Wang, G.X.; Zhan, R.B. A new species of middle Rhuddanian Halysites (Tabulata) from Meitan, northern Guizhou, Southwest China. *Est. J. Earth Sci.* **2015**, *64*, 105–109. [CrossRef]
17. Rong, J.Y.; Wang, Y.; Zhan, R.B.; Fan, J.X.; Huang, B.; Tang, P.; Li, Y.; Zhang, X.L.; Wu, R.C.; Wang, G.X.; et al. Silurian integrative stratigraphy and timescale of China. *Sci. China Earth Sci.* **2019**, *62*, 89–111. [CrossRef]
18. Li, Z.M.; Liu, G.X. New Halysitida from middle Silurian in north-west Hubei and south Shanxi. *Earth Sci.* **1988**, *15*, 481–486.
19. Wang, G.X.; Ian, G.P.; Yong Yi, Z. The youngest Ordovician (latest Katian) coral fauna from eastern Australia, in the uppermost Malachis Hill Formation of central New South Wales. *Alcheringa Australas. J. Palaeontol.* **2020**, *44*, 356–378. [CrossRef]
20. Deng, Z.Q.; Wang, G.X. The Llandoverian coral fauna from xikeer, western tarim basin. *Acta Palaeontol. Sin.* **2012**, *51*, 176–185. (In Chinese, English Summary) [CrossRef]
21. Huang, Z.X.; Cheng, L.R. The Devonian and Silurian corals from Bange and their and Shenzha, northern Xizang. *J. Chang. Univ. Earth Sci.* **1988**, *3*, 249–257. (In Chinese)
22. Li, Y.X. Silurian and Devonian Tabulate coral assemblages in the Beishan area, Gansu Province. *Acta Palaeontol. Sin.* **1983**, *22*, 73–143. (In Chinese, English Summary)

23. Lin, B.Y.; Wang, B.Y. Middle and Late Silurian Tabulata and Heliolitida from the northern side of Bolhinur Mountains in Xinjiang. *Xinjiang Geol.* **1984**, *1*, 37–49. (In Chinese, English Summary)
24. Wang, B.Y. Silurian faunal and paleogeographical characters of the Tianshan Mountains of Xinjiang. *Xinjiang Geol.* **1988**, *6*, 4. (In Chinese, English Summary)
25. Hou, F.H.; Fang, S.X.; Wang, Z.Y.; Zhang, T.S.; Lan, G.; Deng, M.; Cai, Y.L. Organic reef in Lower Silurian. Guangyuan, Sichuan to Ningqiang, Shaanxi. *Acta Sedimentol. Sin.* **1994**, *12*, 7. (In Chinese, English Summary)
26. Li, C.; Wu, X.J.; Fan, J.X.; Chen, Q.; Li, G.; Sun, Z.Y.; Zhang, Y.D. Carbon isotope chemostratigraphy of the Ordovician-Silurian transition interval of the Xike-1 drillcore in Guizhou, China. *Geochemistry* **2019**, *48*, 11. (In Chinese, English Summary)
27. Deng, Z.Q. Some palaeozoic Tabulate corals from northern Xinjiang. In *Palaeozoic Fossils of Northern Xinjiang*; Academia Sinica; Nanjing University Press: Nanjing, China, 1999; pp. 187–269. (In Chinese, English Summary)
28. Wang, G.X.; Zhan, R.B.; Deng, Z.Q. Latest Ordovician and earliest Silurian Tabulate corals of South China. *GFF* **2015**, *136*, 290–293. [CrossRef]
29. Hu, Z.J.; Gong, L.Z.; Yang, S.W.; Wang, H.D. New Knowledge of the Ordovician Silurian Boundary Strata in Shiqian, Guizhou. *J. Stratigr.* **1983**, *7*, 140–142. (In Chinese, English Summary)
30. Mao, Y.Y.; Li, Y.; Ni, C. Bio and lithofacies differentiations of the Leijiatun Formation (late Areonian, Llandovery, Siliurian) in Shiqian, NE Guizhou, SW China. *J. Micropaleontol.* **2015**, *32*, 12. (In Chinese, English Summary) [CrossRef]
31. Liang, K.; Elias, R.J.; Lee, D.J. Morphometrics, growth characteristics, and phylogenetic implications of *Halysites* catenularius (Tabulata, Silurian, Estonia). *J. Paleontol.* **2018**, *93*, 215–231. [CrossRef]
32. Golding, R.E.; Jones, A.S. Micro-CT as a novel technique for 3D reconstruction of molluscan anatomy. *Molluscan Res.* **2007**, *27*, 123–128. [CrossRef]
33. Ren, J.J. Application of Computed Tomography and Other Techniques in Gaojiashan Lagerstätte. Ph.D. Thesis, Northwestern University, Evanston, IL, USA, 2019. (In Chinese, English Summary)
34. Li, J.H.; Pei, R.; Teng, F.F.; Qiu, H.; Tagle, R.; Yan, Q.Q.; Wang, Q.; Chu, X.L.; Xu, X. Micro-XRF study of the troodontid dinosaur Jianianhualong tengi reveals new biological and taphonomical signals. *At. Spectrosc.* **2020**, *42*, 1–11. [CrossRef]
35. Liang, Y.; Zhang, Z.F. Exploration of Cambrian fossils by micro-x ray fluorescence spectrometer. *Acta Palaeontol. Sin.* **2018**, *57*, 10. (In Chinese, English Summary)
36. Luo, L.Q.; Zhan, X.C.; Li, G.G. *X-ray Fluorescence Spectrometer (Analytical Instrument Usage and Maintenance Series) (Precision)*; Chemical Industry Press: Beijing, China, 2008. (In Chinese)
37. Saleh, F.; Pittet, B.; Sansjofre, P.; Guériau, P.; Lalonde, S.; Perrillat, J.P.; Vidal, M.; Lucas, V.; El Hariri, K.; Kouraiss, K.; et al. Taphonomic pathway of exceptionally preserved fossils in the Lower Ordovician of Morocco. *Geobios Paleontol. Stratigr. Paleoecol.* **2020**, *60*, 99–115. [CrossRef]
38. Zhang, Z.F.; Strotz, L.C.; Topper, T.P.; Chen, F.Y.; Chen, Y.L.; Liang, Y.; Zhang, Z.L.; Skovsted, C.B.; Brock, G.A. An encrusting kleptoparasite-host interaction from the early Cambrian. *Nat. Commun.* **2020**, *11*, 2625. [CrossRef]
39. Walczak, M.; Tarsińska-Petruk, D.; Płotek, M.; Goryl, M.; Kruk, M.P. MA-XRF study of 15th–17th century icons from the collection of the National Museum in Krakow, Poland. *X-ray Spectrom.* **2018**, *48*, 303–310. [CrossRef]
40. Schrøder, A.E.; Wielandt, D.K.P.; Rasmussen, J.A.; Carnevale, G.; Storey, M. Benchtop micro–X-ray fluorescence, µXRF: An exciting tool for anatomical studies of fossil bony fishes. *Lethaia* **2023**, *56*, 1–29. [CrossRef]
41. Li, D.; Tan, L.; Guo, F.; Cai, Y.; Sun, Y.; Xue, G.; Cheng, X.; Yan, H.; Cheng, H.; Edwards, R.L.; et al. Application of Avaatech X-ray fluorescence core-scanning in Sr/Ca analysis of speleothems. *Sci. China Earth Sci.* **2019**, *62*, 964–973. [CrossRef]
42. Leitão, R.G.; Silva, M.P.; Diniz, M.S.; Guerra, M. Mapping the distribution of mercury (II) chloride in zebrafish organs by benchtop micro-energy dispersive X-ray fluorescence: A proof of concept. *J. Trace Elem. Med. Biol.* **2022**, *69*, 126874. [CrossRef]
43. Steven, S.; Christina, C.; Eduardo, L.-D. Macro X-ray fluorescence scanning (MA-XRF) as tool in the authentication of paintings. *Microchem. J.* **2018**, *137*, 139–147. [CrossRef]
44. Liang, K.; Elias, R.J.; Lee, D.J. The early record of halysitid Tabulate corals, and morphometrics of Catenipora from the Ordovician of north-central China. *Pap. Palaeontol.* **2018**, *4*, 363–379. [CrossRef]
45. Foote, M.; Miller, A.I. *Principles of Paleontology*, 3rd ed.; W.H. Freeman: New York, NY, USA, 2007.
46. Zhang, Z.L.; Zhang, Z.F.; Holmer, L. Studies on the shell ultrastructure and ontonogy of the oldest Acrotretid brachiopods from south China. *Acta Palaeontol. Sin.* **2017**, *36*, 21. (In Chinese, English Summary) [CrossRef]
47. Berkowski, B.; Zapalski, M.K. Large dwellers of the Silurian *Halysites* biostrome: Rhizosessile life strategies of cystiphyllid rugose corals from the Llandovery of Gotland. *Lethaia* **2018**, *51*, 581–595. [CrossRef]
48. Lee, D.J.; Noble, J.P.A. Colony development and formation in halysitid corals. *Lethaia* **1990**, *23*, 179–193. [CrossRef]
49. Wang, G.X.; Deng, Z.Q. Application of the cluster analysis to classification of Cateniporids. *Acta Palaeontol. Sin.* **2010**, *4*, 9. (In Chinese, English Summary) [CrossRef]
50. Lee, D.J.; Elias, R.J. Mode of growth and life-history strategies of a late Ordovician halysitid coral. *J. Paleontol.* **1991**, *65*, 191–199. [CrossRef]
51. Hammer, Ø. Computer-aided study of growth patterns in tabulate corals, exemplified by Catenipora heintzi from Ringerike, Oslo Region. *Nor. Geol. Tidsskr.* **1999**, *79*, 219–226. [CrossRef]
52. Hammer, O. Regulation of astogeny in halysitid tabulates. *Acta Palaeontol. Pol.* **1998**, *43*, 635–651.

53. Webby, B.D. Succession of Ordovician coral and stromatoporoid faunas from central-western New South Wales. In *International Paleontological Symposium on the Study of Fossil Corals Novosibirsk, Russia, Drevnie Cnidaria II*; Sokolov, B.S., Ed.; Trudy Instituta geologii i geofiziki: Moskva, Russia, 1975; pp. 57–67.
54. Liu, F.; Skovsted, C.B.; Topper, T.P.; Zhang, Z.F. Soft part preservation in hyolithids from the lower Cambrian (Stage 4) Guanshan Biota of South China and its implications. *Palaeogeogr. Palaeoclimatol. Palaeoecol.* **2020**, *562*, 110079. [CrossRef]
55. Hu, Y.Z.; Strotz, L.C.; Knaust, D.; Wang, J.; Liang, Y.; Zhang, F. Distinguishing borings and burrows in intraclasts. *Est. J. Earth Sci.* **2023**, *443*, 106302. [CrossRef]
56. Ou, Q.; Han, J.; Zhang, Z.F.; Hu, D.; Mayer, G. Three Cambrian fossils assembled into an extinct body plan of cnidarian affinity. *Proc. Natl. Acad. Sci. USA* **2017**, *114*, 8835. [CrossRef]
57. Duan, X.; Li, Z.; Liu, M.; Zou, Y.C. Progress of the Iron-mediated soil organic carbon preservation and mineralization. *Adv. Earth Sci.* **2022**, *37*, 202–211. [CrossRef]
58. Adachi, N.; Yoichi, E.; Jianbo, L. Early Ordovician shift in reef construction from microbial to metazoan reefs. *PALAIOS* **2011**, *26*, 106–114. [CrossRef]
59. Xue, Y.S.; Tang, T.F.; Yu, C.L. Discovery of oldest skeletal fossils from upper Simian Doushantuo Formation in Weng'an, Guizhou, and its significance. *Acta Palaeontol. Sin.* **1992**, *31*, 10. [CrossRef]
60. Li, Y.X. The reef facies and mixed facies lower Silurian of Shaanxi, Gansu and Ningxia districts. *Geol. Rev.* **1980**, *26*, 479–485.
61. Wood, R.A.; Grotzinger, J.P.; Dickson, J.A.D. Proterozoic Modular Biomineralized Metazoan from the Nama Group, Namibia. *Science* **2002**, *296*, 2383–2386. [CrossRef]
62. Scrutton, C.T. The Palaeozoic corals, I: Origins and relationships. *Proc. Yorks. Geol. Soc.* **1997**, *51*, 177–208. [CrossRef]
63. Liu, P.J.; Yin, C.Y.; Tang, F. Fossils of microtubule shaped metazoans with multiple bifurcations in the Weng'an biota. *Chin. Sci. Bull.* **2006**, *51*, 488–490. [CrossRef]
64. Chen, J.Y.; Oliveri, P.; Gao, F.; Dornbos, Q.S.; Li, C.W.; Bottjer, D.J.; Davidson, E.H. Precambrian Animal Life: Probable Developmental and Adult Cnidarian Forms from Southwest China. *Dev. Biol.* **2002**, *248*, 182–196. [CrossRef] [PubMed]
65. Xiao, S.H.; Yuan, X.L.; Andrew, H.K. Eumetazoan fossils in terminal Proterozoic phosphorites? *Proc. Natl. Acad. Sci. USA* **2000**, *97*, 13684–13689. [CrossRef] [PubMed]
66. Park, T.Y.; Woo, J.; Lee, D.J.; Lee, D.C.; Lee, S.B.; Han, Z.Z.; Chough, S.K.; Choi, D.K. A stem-group cnidarian described from the mid-Cambrian of China and its significance for cnidarian evolution. *Nat. Commun.* **2011**, *2*, 442. [CrossRef]
67. Jean, L.; Françoise, D.; Anna, G.; David, G. The oldest tabulate coral and the associated archaeocyatha, Lower Cambrian, Flinders Ranges, South Australia. *Geobios* **1991**, *24*, 697–718. [CrossRef]

life

MDPI

Article

The Palaeobiology of Two Crown Group Cnidarians: *Haootia quadriformis* and *Mamsetia manunis* gen. et sp. nov. from the Ediacaran of Newfoundland, Canada

D. McIlroy [1,*], G. Pasinetti [1], D. Pérez-Pinedo [1], C. McKean [1], S. C. Dufour [2], J. J. Matthews [3], L. R. Menon [4], R. Nicholls [5] and R. S. Taylor [1]

[1] Department of Earth Sciences, Memorial University of Newfoundland, St. John's, NL A1C 5S7, Canada; gpasinetti@mun.ca (G.P.); dperezpinedo@mun.ca (D.P.-P.); cmckean@mun.ca (C.M.); rodt@mun.ca (R.S.T.)
[2] Department of Biology, Memorial University of Newfoundland, St. John's, NL A1C 5S7, Canada; sdufour@mun.ca
[3] Museum of Natural History, University of Oxford, Oxford OX3 7DQ, UK; jack.matthews@oum.ox.ac.uk
[4] Department of Earth Sciences, University of Oxford, Oxford OX1 3AN, UK; menon891@btinternet.com
[5] BobNichollsArt, Bristol BS161QY, UK; bobnichollsart@gmail.com
* Correspondence: dmcilroy@mun.ca

Abstract: The Ediacaran of eastern Newfoundland preserves the world's oldest known eumetazoan body fossils, as well as the earliest known record of fossilized muscular tissue. Re-examination of the holotype of the eight-armed *Haootia quadriformis* in terms of its morphology, the arrangement of its muscle filament bundles, and hitherto undescribed aspects of its anatomy support its interpretation as a crown staurozoan. We also document several new fossils preserving muscle tissue with a different muscular architecture to *Haootia*, but with only four arms. This new material allows us to describe a new crown group staurozoan, *Mamsetia manunis* gen. et sp. nov. This work confirms the presence of crown group medusozoan cnidarians of the Staurozoa in the Ediacaran of Newfoundland circa 565 Ma.

Keywords: Ediacaran; Cnidaria; fossil; taphonomy; biostratinomy; Staurozoa; proterozoic

1. Introduction

During the Ediacaran period, Newfoundland lay on the eastern margin of Gondwana and was part of the Avalonian Terrane [1–3]. The most significant finds of Ediacaran fossils in Avalonia come from the Charnian Supergroup of Leicestershire, UK [4–14] and the Conception and St. John's groups across southeastern Newfoundland [15–30]. Both regions are characterized by volcaniclastic successions deposited in deep marine settings [31–34]—shallowing to fossiliferous offshore shelf and prodelta settings in Newfoundland [35]—and have fossil assemblages dominated by rangeomorph fossils [23,36,37].

The Ediacaran of Newfoundland consists of three distinct biotas: (1) those of the Mistaken Point type on the Avalon Peninsula [17,18,38,39]; (2) the exceptionally preserved Upper Island Cove assemblage [22,25,31,40]; and (3) the assemblages of the Catalina Dome on the Bonavista Peninsula [24,29,30,33,41–43] (Figure 1).

The shortage of geochronological data from the Bonavista Peninsula precludes reliable correlation with the now well-dated stratigraphy of the Avalon Peninsula assemblages [44,45], owing to the unreliability of lithostratigraphy as a correlative tool over hundreds of kilometres.

The holotype of the oldest known cnidarian, *Haootia quadriformis* (Figure 2A), comes from the Fermeuse Formation, at Back Cove, near Port Union on the Bonavista Peninsula [29,42] (Figure 1C,D). The type specimen was later removed from the field with permission of the Province of Newfoundland and Labrador, and is now housed in The Rooms in St. John's Newfoundland (NFM F-994). The monotypic genus is based on plastotype OUM ÁT.424/p at the Oxford University Museum of Natural History. The well-preserved holotype and the paratype both include the preservation of muscle fibres, which have been used to interpret *Haootia* as a probable cnidarian [29,42,46,47].

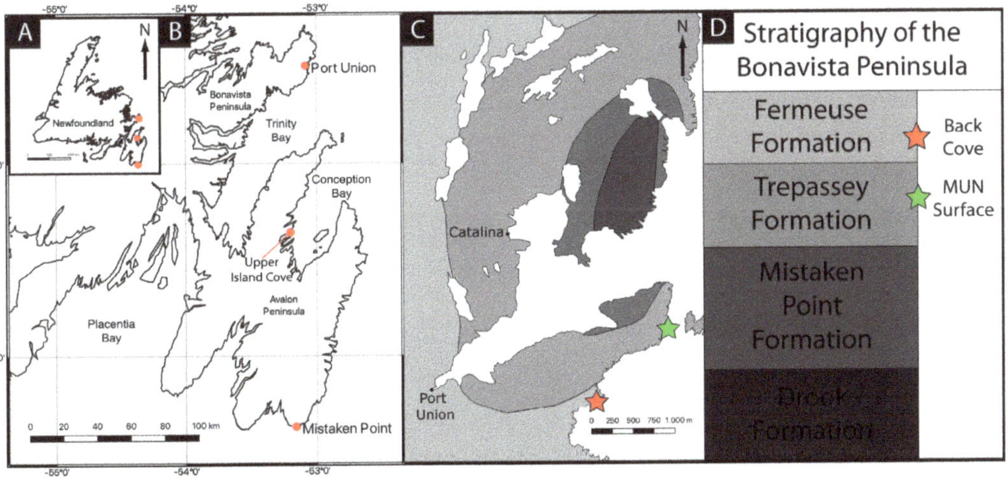

Figure 1. Location of the Catalina Dome on the Bonavista Peninsula of the island of Newfoundland, Canada, with key fossil sites highlighted with red circles (**A**,**B**) and the fossil sites with *Haootia* and *Mamsetia* highlighted by red and green stars, respectively (**C**,**D**). *Mamsetia* lies some 150 m stratigraphically below the type locality of *Haootia*.

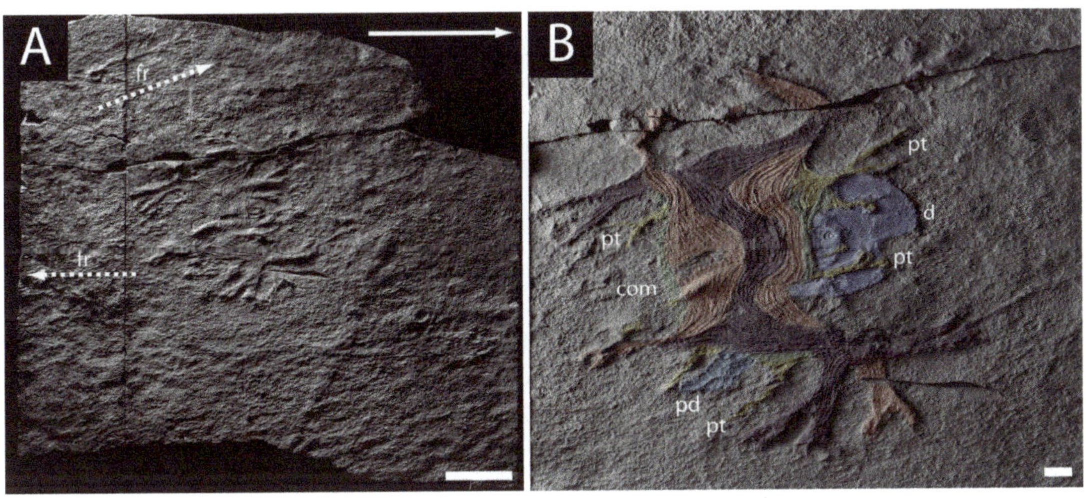

Figure 2. Holotype specimen of *Haootia quadriformis*. (**A**) image of the complete slab housed at The Rooms Provincial Museum, St. Johns, NL, with the inferred palaeocurrent indicated with a white arrow, scale bar 5 cm; (**B**) detail of the same specimen with groups of muscle fibres highlighted in colour: (pt) primary tentacle, (ped) peduncle, (com) coronal muscle, (d) disc; scale bar is 1 cm.

Our re-examination of *Haootia* is based on the type material, including additional morphological and taphonomic considerations, as well as new, comparably preserved material attributed to a new genus and species of staurozoan cnidarian. We focus on the detailed description of the muscular tissue of the calyx and arms, and the relationship to the previously inferred basal disc [29,42]. The importance of *Haootia*—and our new genus

Mamsetia—is emphasised by recent progress in understanding relationships within the Cnidaria [14,48,49].

2. Fossil Evidence for the Origins of the Cnidaria

The early fossil record of Cnidaria is largely that of the Medusozoa and Anthozoa, fossil examples which have been documented as present in a number of late Ediacaran sites [14,27,47,50]. Early interpretations of the frondose components of the Ediacaran macrobiotas typically compared most of the frondose taxa with pennatulacean anthozoans [51], and most discoidal taxa with medusozoans [52]. Those historical ideas have largely been replaced as many key taxa—particularly the Rangeomorpha and Arboreomorpha—are uniquely constructed and quite dissimilar to the Anthozoa [17,53]. Many of the Ediacaran fronds are composed of numerous self-similar modules in a fractal-like manner [7,10,19,20,22,25,26,37,54–56]. The discovery that many of the circular Ediacaran fossils originally interpreted as medusoids are either holdfast discs of frondose taxa (see reviews in [57–59], but also see [50]), or pseudofossils [27,60,61] has greatly decreased the number of inferred cnidarians in the Ediacaran.

Evidence for possible cnidarians from the late Ediacaran to Cambrian includes: (1) trace fossil evidence [27,62–64] of some un/lightly mineralized tubular taxa that have been compared to scyphopolyps (e.g., *Corumbella* [65]; and *Wutubus* [66]); (2) the conulariids *Vendoconularia* and *Paraconularia* [47,67–70], and the tetra-radial tubicolous medusozoan polyp *Auroralumina* [14]; and (3) the preservation of soft-bodied organisms in Cambrian lagerstätte [71,72]. In addition, many of the tubicolous taxa compared to the cnidaria are only known from the latest Ediacaran to Cambrian, making the cnidarian affinities of *Haootia quadriformis* all the more important to determine.

3. Taphonomy/Biostratinomy of *Haootia quadriformis*

Fossiliferous Ediacaran surfaces in the Catalina Dome (Bonavista Peninsula, Newfoundland) preserve exquisitely fine details of fossils in combinations of negative and positive epirelief [30], sometimes also involving a ferruginous veneer [43]. A combination of chemical and mechanical (salt) weathering affects the bedding surfaces by removing overlying ash layers and exposing the underlying fossils [73]. The main natural sources of damage to fossils consist of falling rocks, ice wedging, percussion impact of pebbles during storms, and scour from sea-ice [74].

Haootia is distinctive among the Ediacaran fossils of Newfoundland in that it preserves muscle fibres (Figure 3A,B) [29,42]. The calyx of *Haootia* is preserved slightly above the surrounding bedding plane in a net-positive epirelief in the manner of some rangeomorphs (e.g., *Fractofusus* [56] and *Beothukis* [75]), although the fine details are generally negative-relief features on the (net) positive-relief fossil (Figure 3A,B,D).

Palaeocurrent data from frond orientations in the Catalina Dome show trends predominantly towards the southeast [24,76], which is consistent with independent current indicators [77]. The paradigm of frond "felling" has been challenged for some Ediacaran taxa [75], though some taxa—particularly some of the frondose arboreomorphs such as *Arborea spinosa* [78] and some rangeomorphs including *Avalofractus* and *Primocandelabrum*—do seem to have been erect during life [17,25,37,77–79]. Importantly, however, the erect fronds were likely oriented relative to clear-water currents, rather than density currents [77,79].

While sediment smothering by volcaniclastic material (either from a density current or ashfall settled from suspension [19,45,75]) is the most commonly invoked mode of preservation of the Ediacaran assemblages of Newfoundland, the preservation of *Haootia* was previously attributed to entrainment [42]. This would be comparable to the mode of preservation originally posited for the Upper Island Cove biota [25], but subsequently refuted [31,40]. Since the type material of *Haootia* is on a bedding plane, not within a bed, it was likely smothered in an obrution type event [80–84], not entrained within a sediment-laden current. The preservation of most of the arms on the inferred up-current side of the fossil (Figure 2B) supports the idea that the *Haootia* organism was affected by a unidirectional current prior to smothering. The calyx of *Haootia* should have forced the organism to be oriented down-current of a holdfast—if indeed one was present—but since this is not the case, we propose instead that the organism held onto the seafloor using the arms and tentacles on the up-current side of the calyx (see discussion of tentacle function in [85]). This is similar to the in-life position commonly assumed by modern staurozoans affected by currents [86]. The mode of preservation is thus most likely to have involved (tuffaceous) sediment obtrusion, and very early diagenesis [83].

The preserved morphology of *Haootia* is inferred as resulting from the preservation of muscle tissue, specifically bundles of muscle fibres [29]. Lineations within the musculature of *Haootia* are mostly concave epireliefs, implying that they represent collapsed muscle fibres (Figure 3A,B) rather than the permineralization or replacement of muscle tissue. Collapse of the muscle tissues must have happened after early lithification of the overlying ash, eventually allowing the underlying silt to fill the void left by the body [87,88]. The remobilization of silt likely happened only after degradation of the associated microbial mat during early burial [89]. Associated rangeomorph fossils on the MUN surface usually preserve the lower surface of the organism at or below the ambient bedding plane [30], which is apart from the normal positive relief preservation of some stems [90].

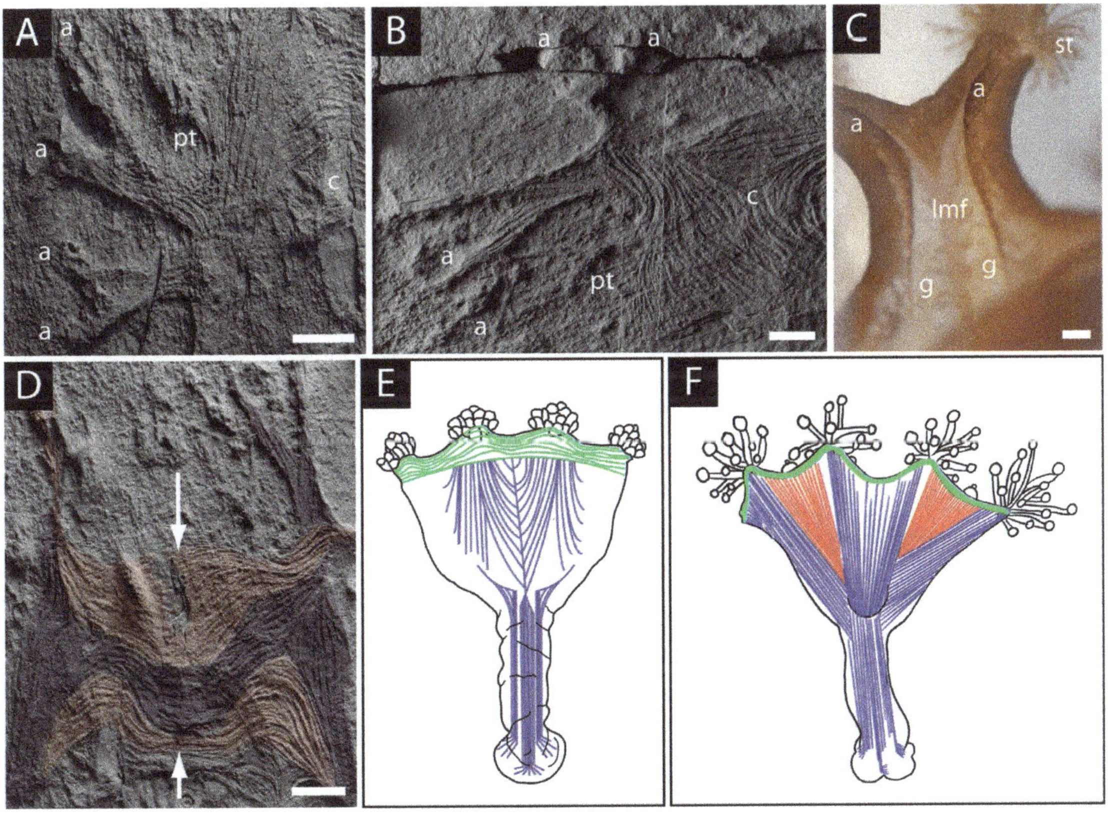

Figure 3. Comparative musculature of *Haootia quadriformis* and the modern staurozoan *Lucernaria*: (**A**) bundles of muscle fibres of branching arms (a) extending into the calyx (c) of the holotype of *H. quadriformis*, which is raised slightly above the bedding plane, scale bar 1 cm; (**B**) detail of the muscle fibres of the arms (a) and calyx (c) of *H. quadriformis*, showing their flattened nature and preservation slightly above the ambient bedding plane; note the twisting and branching of the arm, and also the muscles at the base of the primary tentacle (pt), scale bar 1 cm; (**C**) detail of an arm of

Lucernaria showing the longitudinal muscle fibres (lmf) and branching of the arm (a) (partly covered by gonads (g)), branches of the arm are tipped by secondary tentacles (st), scale bar 0.25 cm; (**D**) groups of muscles of *H. quadriformis* showing the interlacing of one set of muscle fibres with the ends of the fibres from the opposite side of the calyx, forming an irregular vertical seam (arrowed) between the two groups of muscles (orange), and a second set of muscles (purple) meet the muscles of the opposing side of the calyx at its base, scale bar 1 cm; (**E**) musculature of the calyx of the extant staurozoan *Manania* showing the manner in which the longitudinal muscles of the arms (blue) meet to define the shape of the calyx in a manner similar to that seen in *Haootia* (after [91]); (**F**) musculature of the calyx of the extant staurozoan *Haliclystus*, showing the more open nature of the calyx with the longitudinal muscles of the arms (blue) not meeting at the margin of the calyx as inferred for *Mamsetia manunis* gen et sp. nov. (after [91])—perradial musculature (red) is not seen in *Mamsetia*. All scale bars 1 cm.

The muscles of *Haootia* constitute a four-sided calyx with pairs of arms at each corner (Figure 2B). Where groups of muscle fibres cross one another, it is the uppermost set of muscle fibres that is preserved, though often it is possible to determine the position of underlying muscle fibres where they can deform overlying muscle layers (Figure 3A,B). Tissues that lay below the uppermost musculature of the calyx of *Haootia* were not cast by the overlying tuff and were therefore not preserved. The preservation of muscle fibres is relatively common in chordates [92,93], and in some early invertebrates such as Cambrian lobopods [94,95] and Cnidaria [50]. With taphonomic limitations in mind, it is nonetheless clear that the preservational quality of the holotype of *Haootia* is truly exceptional [29,42].

4. Descriptive Palaeontology

Careful photography and tracing of groups of inferred muscle fibres that form the basis of the redescription of the type material of *Haootia* herein support its inclusion in the Cnidaria [29,42,46], and invite comparisons with the tetra-radial bauplan of the Staurozoa as detailed below.

4.1. Longitudinal Muscles of the Calyx and Arms of H. quadriformis

The predominant feature of the *Haootia quadriformis* holotype is the almost-square outline of the calyx [29]. The calyx is composed of eight sets of longitudinal muscles, all of which extend beyond the calyx into an arm of the fossil (Figure 2B). The more completely preserved of the arms bifurcate close to their tips (Figure 2B). Preserved morphology of the tips of the arms is completely lacking in the holotype, suggesting that the tissues were rapidly degraded, and thus difficult to fossilize.

The arms of modern stauromedusae have clusters of secondary tentacles at their tips (cf. Figure 3C) that are used for grasping food particles, including small invertebrates [96]. In the laboratory, the secondary tentacles of the arms of *Lucernaria* are commonly seen to brush the sediment surface (Supplementary Information S1), which suggests possible surficial deposit feeding. The arms of Staurozoa are also important in some species for locomotion on both hard and soft substrates, usually involving initial attachment using the secondary tentacles (at the ends of the arms) and (primary tentacle) anchors, followed by contraction of the longitudinal muscles of the arms [97] (Supplementary Information S2).

While no tentacular structures are preserved at the ends of the arms of *Haootia*, it is possible, given the gross morphological similarity with the Staurozoa, that there were clusters of secondary tentacles at the tips of at least some of the arms. The preservation potential of delicate cnidarian tentacles is extremely low. Even in konservat lagerstätte, in which muscle tissues are moderately well preserved, fine tentacles are generally poorly preserved [59,98,99]. More robust cnidarian tentacular structures have recently been inferred from the Ediacaran *Auroralumina* ([14] though no muscle preserved) and the Cambrian *Burgessomedusa* [100].

The corners of the calyx in the holotype of *Haootia* are characterized by pairs of arms (Figure 2B). The musculature of one set of arms—one of which is present in each corner of the calyx—is arranged such that it meets the musculature of longitudinal muscles from the opposite side of the calyx at its base (Figure 2B, coloured purple). The other set of arms have fibres that are more curved in the calyx portion, and connect to the equivalent muscle fibres of the adjacent paired arms (Figures 2B and 3D, coloured orange), as in some modern staurozoans (Figure 3E).

In absence of preserved tissues from the oral surface of the arms (e.g., gonads and gastric tissues, which would be highly labile and have low fossilization potential, Figure 3C), it is possible that some of the arms were specialized for locomotion rather than feeding (Figure 4).

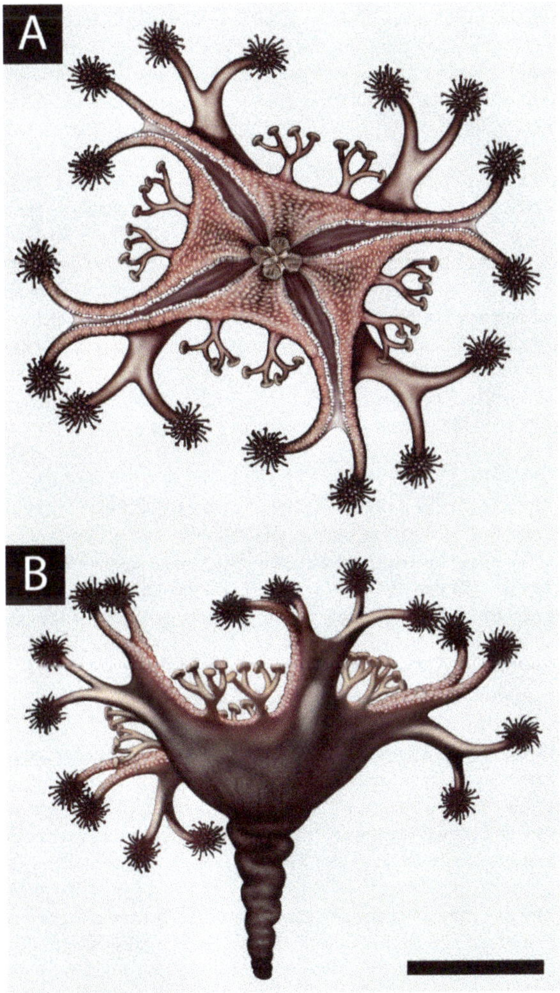

Figure 4. The holotype of *H. quadriformis* reconstructed by analogy with modern stauromedusae, showing the paired arms at each corner of the calyx tipped with hypothetical secondary tentacles and shorter divided primary tentacles at the margin of the calyx with hypothetical adhesive pads inferred from the morphology of modern stauromedusae. Approximate scale bar 5 cm. (**A**) is an oral view and (**B**) is the lateral view of the calyx and peduncle © Bob Nicholls.

We note that a free-living developmental stage of the parasitic cnidarian *Polypodium* [101,102] has four walking tentacles and four feeding tentacles, all with longitudinal muscles [102,103]—an architecture somewhat comparable to *Haootia*. *Polypodium* might, therefore, be a good supplementary biomechanical model to understand the eight-armed *Haootia*. No taxonomic affinity is inferred, however, as *Polypodium* is probably a highly derived cnidarian.

4.2. Marginal Muscles of the Calyx of H. quadriformis

Our re-examination of the holotype *Haootia* has demonstrated the presence of low-relief bands of narrow muscle fibres that lie parallel to the margins of the inferred calyx (Figure 5A,B), which are consistent with *Haootia* having a stauromedusan-like calyx [91]. The coronal (marginal) muscles of Staurozoa form a similar narrow band on the margin of the calyx (Figure 5C) and serve the function of closing it, for example during times of environmental stress [91] or to decrease the aspect ratio of the calyx in a strong current [96]. The coronal muscles of *Haootia* do not continue across the arms, suggesting that they were either discontinuous, as they are in many staurozoans, or that they ran inside the arms of the holotype and are thus not preserved because of preferential casting of the overlying arms.

The poorly developed marginal muscles of the Staurozoa are in marked contrast with those of free-swimming medusozoans (e.g., Scyphozoa and Cubozoa), in which contractions of the marginal/circular muscles are used for propulsion [104]. The paucity of marginal musculature in stauromedusae is a function of their sessile epibenthic mode of life [105]. It has previously been stated that the coronal musculature of *Haootia quadriformis* was extremely well developed, and more comparable to that of pelagic cnidarians [46]. Those assertions are based on misinterpretation of some of the longitudinal muscles of the arms that lie parallel to the margin of the calyx circular/coronal muscle tissue. That same misinterpretation of some of the longitudinal musculature of *Haootia* as coronal has previously cast some doubt on the interpretation of *Haootia* as a crown group cnidarian [14,46,69].

4.3. Marginal Tentacles of H. quadriformis

The holotype of *Haootia* has short tentacles associated with the margin of the calyx [29] (Figures 2B and 5A,B). The presence of "small branching structures" was noted in the diagnosis of *Haootia* [29], but they were not recognized as being analogous to the primary tentacles of Staurozoa as they are herein. Among the Cnidaria, similarly positioned tentacles are common in stauropolyps, but in the stauromedusae these primary tentacles may be: (1) resorbed [106]; (2) metamorphosed into inter-radial anchors [107]; (3) modified in situ at their juvenile position [106]; or (4) become capitate and migrate into clusters with the secondary tentacles at the end of the arms [107]. When stauromedusan primary tentacles are located on the calyx margin, they are intimately associated with the coronal muscle, and are controlled by perradial musculature [91]. The marginal tentacles on the coronal/marginal muscles of the calyx of *Haootia* all have well-developed musculature (Figure 2A). The distal ends of the marginal tentacles appear to be filiform, although the preservation is only of the internal musculature, not of the gross morphology of the original structure (Figure 5A).

Modern stauromedusae do not generally have pairs of perradial primary tentacles, although many stauropolyps have eight primary tentacles in the early phases of growth [108], four of which are either resorbed or migrate to the ends of the arms [107]. Rare examples of *Halyclistus* have been documented with paired perradial anchors (e.g., [106], their Figure 7j), although whether this phenotype is associated with genetic or environmental drivers is unclear (see [109]).

Among modern Staurozoa, the amyostaurid *Calvadosia* has long, prominent, primary tentacles, similar to those of *Haootia*, with a distinctive V-shaped region of perradial muscle that extends beyond the marginal muscle in a manner that is comparable to the musculature at the base of the marginal tentacles in *Haootia* (Figure 5A). The long primary tentacles of *Calvadosia* are primarily used to modify the position of the calyx by attaching the calyx to the substrate (e.g., during locomotion [97]). In the holotype of *Haootia*, it seems that one of the marginal tentacles has lifted a portion of the margin of a large holdfast structure, revealing the sub-disc sedimentary texture underneath, suggesting a similar grasping function and the facility to contract the muscles therein (arrowed on Figure 5A).

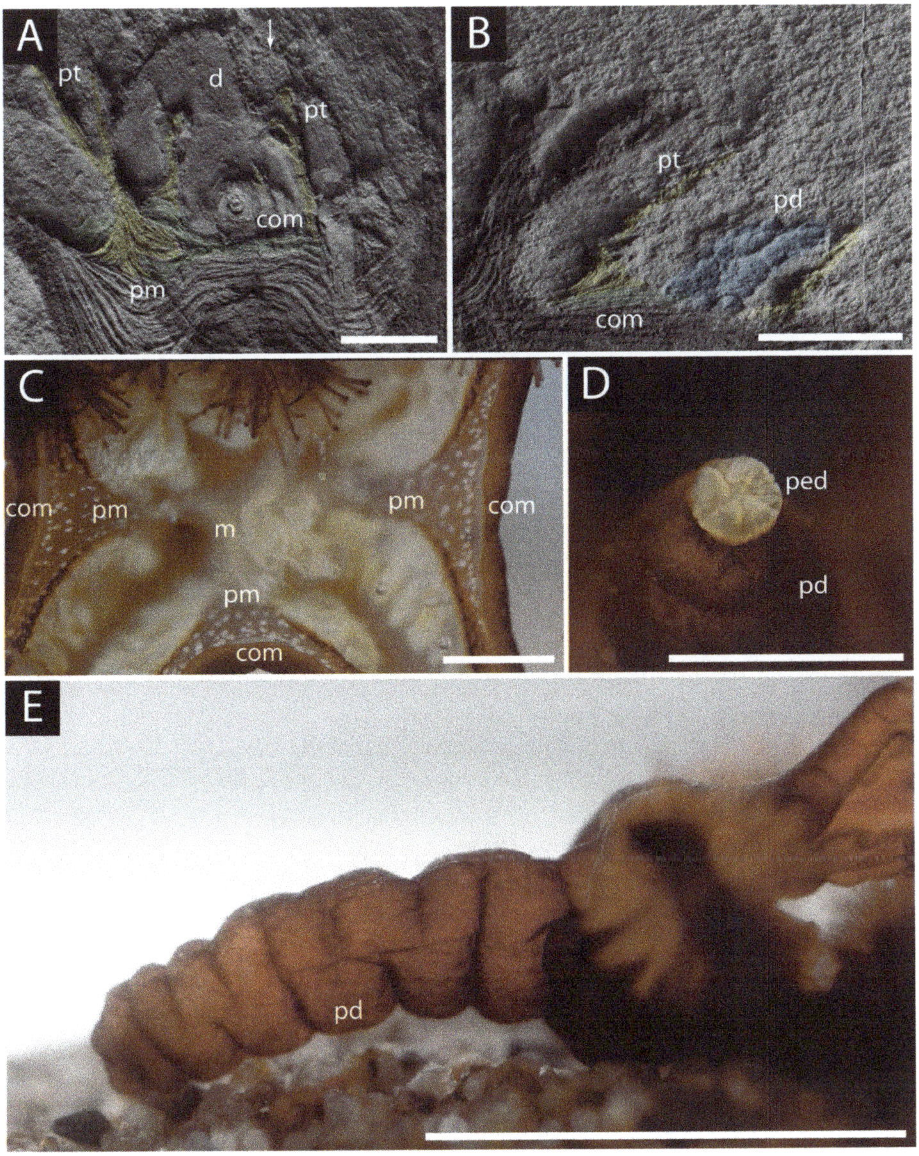

Figure 5. Comparative morphology of *Haootia* and a modern staurozoan (*Lucernaria*): (**A**) primary tentacles (pt) with triangular per-radial musculature (pr) showing their filiform morphology and association with the coronal muscle (com), scale 1 cm; also note the lifting of the margin of a disc (d) by one of the filiform primary tentacles (arrow); (**B**) detail showing the inferred peduncle of *Haootia* (pd) adjacent to a primary tentacle (pt) with coronal muscle (com) (scale 2 cm); (**C**) coronal muscles (com) and per-radial muscles (pm) of modern *Lucernaria* and the central manubrium (m), scale 1 cm; (**D**) pedal disc (ped) of *Lucernaria*, which is small and comparable in width to the peduncle (pd), scale 5 mm; (**E**) morphology of the peduncle (pd) of *Lucernaria*, approx. scale 3 cm.

4.4. Peduncle and Basal Attachment of H. quadriformis

In the description of the holotype [29], the calyx and arms of *Haootia* were borne atop a short stalk or peduncle, which was attached to a large, rather smooth disc (Figure 2B). The paratype of *Haootia* (Figure 6) was also considered to be a partial specimen of a stalk/peduncle and associated disc.

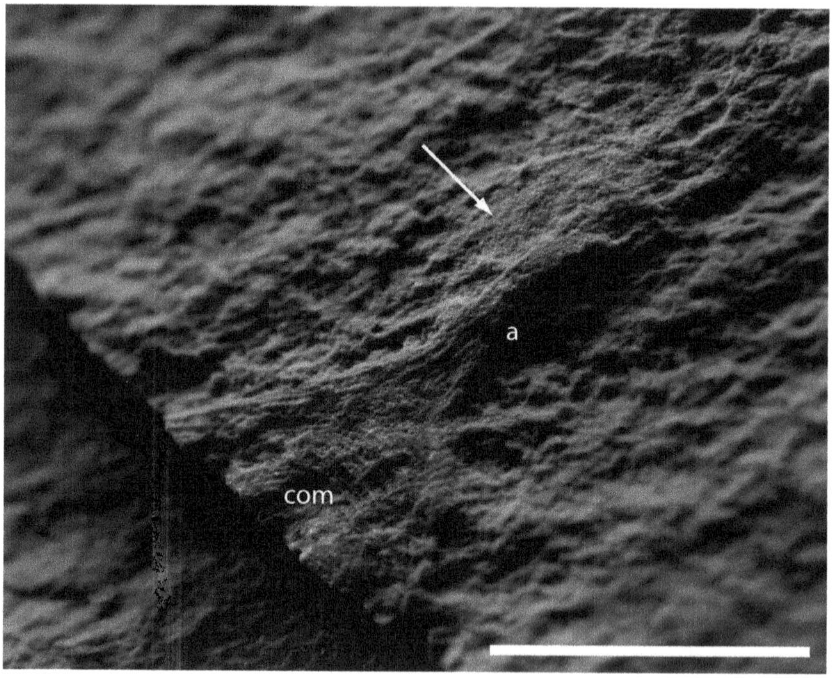

Figure 6. Detail of the specimen formerly designated as the paratype of *H. quadriformis* by [29] and compared to *Mamsetia manunis* gen. et sp. nov. herein: longitudinal muscle of the arm (a) is highlighted in orange—note that there is only one arm at the corner of the calyx, not 2 as in the holotype of *H. quadriformis*. The preserved portion of circular/coronal muscle (com) of the oral surface is more complete than in stauromedusae, but comparable to the oral surface of stauropolyps. Note the smooth area above the arm (arrowed), which was interpreted as a basal disc previously. Scale bar 1 cm.

Immediately adjacent to the calyx of the holotype is a short, roughly conical, wrinkled structure that has not been discussed in previous studies (Figure 5B). The wrinkled structure is quite unlike any of the described Ediacaran taxa from the Catalina Dome [24] or other Ediacaran deposits in the Newfoundland sections [23,36,37]. The peduncle of many modern staurozoans does, however, have a similar wrinkled outer surface, and it is to this structure that this fossilized tissue is compared (Figure 5E).

The peduncle of many, but not all [91], modern Staurozoa contain longitudinal musculature that allows its contraction, either to retract the calyx towards the substrate or to allow locomotion [98,106]. Modern Staurozoa are generally considered to locomote by alternate adhesion of the arms and peduncle to the substrate, combined with contraction of longitudinal muscles in both structures to produce a tumbling or inchworm-like motion [86]. We note, however, that in laboratory settings, the modern staurozoan *Lucernaria* predominantly locomotes on flat, sandy substrates with the manubrium down, using the arms and adhesive properties of the secondary tentacles at their tips (Supplementary Information S1).

There is no evidence for longitudinal muscles being preserved in the peduncle-like structure of *Haootia*. The peduncle might have also contained muscles, but not as the outermost tissue preserved.

We have not found compelling evidence of any direct biological association between the holotype and a discoidal holdfast-like structure—the affinities of the paratype are discussed below. The disc associated with the holotype of *Haootia* is much larger than the pedal disc of modern Staurozoa (Figure 5D), which use the pedal disc for temporary attachment. All known stauromedusae are motile, a mode of life that would be incompatible with a pedal disc that is half the width of the calyx and approximately ten times wider than the peduncle itself [46]. In the absence of reliable evidence linking *Haootia* with a holdfast-like disc, it becomes pertinent to consider a mobile epibenthic mode of life.

4.5. Issues with the Paratype of Haootia quadriformis

The partial specimen described as the paratype of *Haootia* [29] is a field photograph of a small partial specimen of just over 2 cm in length from the MUN surface [42]. It has been considered to have an associated disc of very low relief that is approximately 5 mm in diameter (Figure 1f in [29]; Figure 6). The newly accessioned cast of the former paratype of *Haootia* (NFM F-3976) does not show clear evidence of a disc, just a smooth area of adherent tuffite (Figure 6, arrowed).

Additionally, the arrangement of muscle fibres in what has previously been considered a peduncle most closely resembles the longitudinal muscles of the arms of the holotype. The musculature of the calyx at the basal end of the preserved longitudinal muscle is similar to the coronal musculature expected at the margin of a staurozoan-like calyx (Figure 5C). It seems most parsimonious, therefore, to consider that the specimen preserves the circular marginal muscles of the calyx and the longitudinal muscles of one arm rather than a peduncle and part of the calyx. The absence of two arms at the corner of the calyx—the case in the holotype of *H. quadriformis*—is, however, problematic to its inclusion in the genus *Haootia*.

The upper surface of the calyx in the former paratype of *Haootia* is not open, like those of stauromedusae, but is covered by concentric muscles, which are morphologically comparable to the musculature of stauropolyps and also Hydrozoa [104]. We hope to find further specimens of small, *Haootia*-like fossils, but given the presence of just one (unpaired) arm, and the fact that the calyx is closed, it seems unlikely that the specimen belongs in *Haootia*. It could, perhaps, represent an early developmental stage of an undescribed staurozoan or hydrozoan.

4.6. New Fossils with Muscular Preservation from the MUN Surface

We describe herein for the first time several partial specimens of a large four-armed tetra-radial cnidarian-like muscular organism (Figure 7A–D) from the MUN Surface in the Trepassey Formation of the Catalina Dome. Like the specimen that was originally designated as the paratype of *Haootia quadriformis*, all of the new specimens have only one arm on each corner of the calyx, the state in all modern Staurozoa. The new material has a band of well-developed coronal musculature that lies inside the arms (Figure 7A) and includes some very large specimens (Figure 7B,C), all of which are elongated in the direction of the inferred palaeocurrent on the MUN Surface [76]. None of the new specimens is completely preserved; the most commonly preserved tissues are the coronal muscles and the longitudinal muscles of the arms. The tissues of the arms and calyx appear to be less well connected to one another than in the holotype of *H. quadriformis*, perhaps due to the partial decay of connective tissues or original morphology (cf. Figure 3F). In some species of modern staurozoan, the musculature of the arms does not meet that of adjacent arms to form a muscular margin to the calyx, but instead all muscles meet at the base of the calyx [91] (Figure 3F). The calyx is, thus, not defined by the musculature of the arms, so the arms are less likely to remain associated with one another post-mortem once the calyx and other soft tissue decays. The specimens are preserved as positive relief structures, and have thus been subjected to abrasion by rockfall onto the surface, which degrades the specimens. The four arms divide close to their tips in the best-preserved specimens (Figure 7A). The large size of the specimens suggest that the number of arms present is a characteristic of adult organisms, and the arrangement of the musculature of the arms (one in each corner) similarly suggests that the small, former paratype of *H. quadriformis* should, instead, be considered conspecific with this new material (on the same surface), not *Haootia*, as described below.

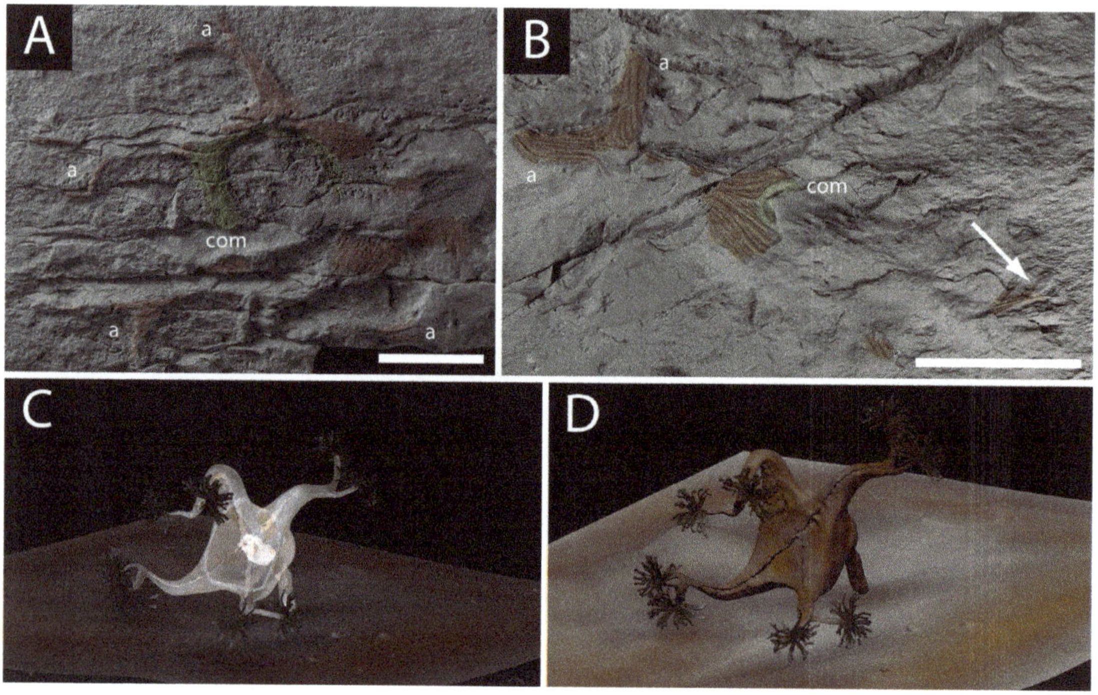

Figure 7. Fossil material and reconstruction of *Mamsetia manunis* gen. et sp. nov.: (**A**) holotype of *M. manunis* preserving a thick band of coronal muscle ((com) coloured green and has no primary tentacles) and longitudinal muscles of the arms ((a) coloured orange)—note that the calyx is not coherent as it is in *H. quadriformis*, and the specimen was damaged by rockfall in the field, scale bar 2.5 cm; (**B**) large specimen of *M. manunis* (scale bar 5 cm), orange longitudinal muscles of the arms, scale bar 5 cm (a), and possible green coronal muscle (com)—note the branching at the tips of the arms (arrowed); (**C**) reconstruction of *Mamsetia manunis* showing the four arms and a staurozoan body plan with no primary tentacles. The exterior tissue of the calyx is made transparent to highlight the positions of the internal coronal and longitudinal musculature; (**D**) reconstruction of the mode of life of *Mamsetia manunis*.

5. Systematic Palaeontology

Phylum: Cnidaria Hatschek 1888
Subphylum: Medusozoa Peterson 1979
Class: Staurozoa Marques and Collins 2004
Genus: *Haootia* Liu et al., 2014
Diagnosis: per species.
Type Species: *quadriformis* by monotypy
Holotype: NFM F-994 (The Rooms Provincial Museum, St. John's, NL, Canada); plastotype OUM ÁT.424/p (Oxford University Museum of Natural History, UK)
2014 *Haootia quadriformis* Liu et al., 2014, figures 1a–e and 2a,

non v2014 *Haootia quadriformis* Liu et al., 2014, figure 1f [moved to *Mamsetia manunis* gen. et sp. nov.]

Emended diagnosis: soft-bodied, tetra-radially-symmetrical fossil with a calyx comprised of numerous long linear fibres that extend into eight arms. The arms divide dichotomously to form smaller sub-branches. The margin of the calyx is marked by a band of circular muscle. The short pedicle structure is not attached to a substantive holdfast or disc. The margin of the calyx has at least two short narrow branches which have muscular tissue at their centre and which divide distally.

Description: the calyx is formed of muscular tissues, as is the case in some modern Staurozoa (e.g., *Manania*, see [91], Figure 3E), with the musculature of adjacent arms meeting to form the margin of the calyx between the arms. The short branches on the margin of the calyx are considered to be staurozoan-like primary tentacles and are preserved in the form of their internal musculature. The contractile nature of the tissue is evidenced by the lifting of an adjacent holdfast disc, possibly post-mortem. The pedicle does not preserve any muscle.

Remarks: the previously inferred disc ([29]) is considered to be an accidental association herein. The type description included the designation of a paratype, which is a very small partial specimen that we consider as having only one arm preserved along with part of the coronal/marginal muscle. The former paratype comes from the same surface as the newly described *Mamsetia manunis* gen. et sp. nov., which similarly has only one arm at each corner of the calyx, and is thus tentatively transferred to that genus herein. While *H. quadriformis* has eight rather than four arms, as is normal in modern Staurozoa, it has sufficient other morphological characters (tetra-radial, peduncle, marginal tentacles) and muscular architecture to include it in that class (Supplementary Information S2).

Genus: *Mamsetia* gen. nov.

v. cf. *Haootia quadriformis* Liu et al., 2014, figure 1f [paratype of *H. quadriformis* Liu et al., 2014].

Diagnosis: per species.

Etymology: *Mamsetia* is derived from "Mamset", which is a word in Beothuk (the language of the indigenous people of Newfoundland at the time of European colonization) meaning 'living'.

Type species: *manunis* sp. nov. by monotypy.

Etymology: The species name *manunis* is derived from "Manune or Manume", which is Beothuk for pitcher/cup, pertaining to the shape of the calyx.

Type Locality: MUN Surface in the Port Union Member of the Trepassey Formation near Green Island, Port Union, Bonavista Peninsula, NL (Figure 1C,D).

Holotype: NFM F-4011 (Figure 7A).

Paratype: NFM F-4012 (Figure 7B).

Other material: NFM F-4013; NFM F-4014; NFM F-4015; NFM F-4016; and possibly NFM F-3976 [former paratype of *Haootia quadriformis*].

Diagnosis: A tetra-radial soft bodied organism comprised of preserved muscle tissue, defining the four corners of the calyx. The arms bifurcate distally into two equal muscular structures. Coronal musculature broad and well-developed, without marginal tentacles. No basal attachment disc is present.

Description: the manner in which the muscles of the arms join to form the calyx is not documented in any of the specimens. The well-preserved coronal muscles of the holotype do not have any evidence of structures comparable to the primary tentacles of some Staurozoa, e.g., *Haliclystus* [91] (Figure 3F), distinguishing it from *H. quadriformis*. None of the specimens of this species shows evidence of a prominent basal disc, as previously invoked for *H. quadriformis* [42].

Remarks: the four arms of *M. manunis* support the attribution of the genus to the Staurozoa, and indirectly support claims that *Haootia* was also a staurozoan. While the beds with both *H. quadriformis* and *M. manunis* have not yet been dated precisely, the best lithostratigraphic estimate of around 565 Ma [42] would make these the oldest crown group

Cnidaria currently known, with *M. manunis* lying in a bed 150 m stratigraphically below that from which *H. quadriformis* was described. In gross morphology at least, *Mamsetia* is considered to have looked much like a modern staurozoan (Figure 7C,D; Supplementary Information S2), with musculature that does not encapsulate the whole calyx but forms bands within it (Figure 7C; compare Figure 3F).

6. Discussion

Mamsetia manunis and *Haootia quadriformis* have several characters typical of stauromedusae, namely: (1) a tetra-radial calyx defined by a total of either four or eight arms, located on the corners of the calyx; (2) the open end of the calyx has a narrow marginal band of coronal/marginal muscle; (3) in *H. quadriformis*, two marginal tentacles are present on each margin of the calyx, each with associated musculature; (4) the arms of *Haootia* have musculature that defines the shape and symmetry of the calyx, much like that of the modern staurozoan *Manania*, whereas that of *Mamsetia* gen. nov. is closer to that of *Haliclystus* [91]; (5) *Haootia quadriformis* preserves a flexible peduncle without permanent attachment to the substrate; and (6) contrary to earlier reports [29], there is no good evidence to support the presence of a broad holdfast disc homologous to the pedal disc of modern staurozoans (Figure 8).

Figure 8. Artist's reconstruction of the seafloor with several *Haootia quadriformis* reconstructed as motile cnidarians with eight arms, marginal tentacles, and a peduncle. The *Haootia* are living in a current (left to right) on a microbe-dominated seafloor alongside a reclining *Charnia* sp. and possible alga *Parviscopa bonavistensis*. Note that one of the muscular primary tentacles of the foreground *Haootia* is lifting the margin of a discoidal holdfast-like form, as in the holotype specimen © Robert Nicholls.

During staurozoan metamorphosis from stauropolyp to stauromedusa, four of the eight primary tentacles present in stauropolyps have a variety of fates: they may be retained into adulthood, while some are resorbed or transformed into either rhopalioids or rhopalia [109,110]. *Haootia quadriformis* retained at least eight marginal tentacles into adulthood (Figure 2A), and *Mamsetia manunis* had none preserved.

Haootia quadriformis (Figure 2) differs from modern stauromedusae in having eight arms that define the tetra-radial symmetry—rather than the four of *Mamsetia* (Figure 7)—but it otherwise closely resembles modern staurozoa. The presence of four additional arms in modern staurozoan polyps opens the possibility of interpreting *H. quadriformis* as an early or "crown" staurozoan, which might have retained the eight arms as a plesiomorphic trait. The preserved features of the newly described *M. manunis* are all present in modern staurozoa, thus supporting the presence of Ediacaran staurozoan cnidarians. The preservational style of *Haootia* and *Mamsetia* is dependent on the casting of muscle tissues [29]. The arms and primary tentacles of *Haootia* only preserve the muscle tissues, with no clear evidence for terminal tentacular structures. If *Haootia* was, indeed, a member of the Cnidaria, it is likely to have either borne nematocysts or adhesive tentacle clusters to aid feeding and/or locomotion (Figure 8).

The subumbrellar surface of the arms in modern Staurozoa commonly bear gonads, but since the delicate structures of the subumbrellar surface (gastrovascular system, manubrium, etc.) of *Haootia* and *Mamsetia* are not preserved, it remains possible that: (1) either all arms were gonad-bearing, like modern stauromedusae (Figure 4); (2) that one set of arms in *H. quadriformis* was purely designed for locomotion (Figure 4); or (3) if all the arms of *Haootia* and *Mamsetia* had no such gastrovascular system, the organism would be less Staurozoa-like than reconstructed herein, but given the tetra-radial symmetry it would likely still be considered a cnidarian. The preserved anatomy of *Haootia* demonstrates that it was a benthic cnidarian of medusozoan affinity, with longitudinal muscles within the calyx and weakly developed marginal/coronal muscles around the mouth of the calyx.

Preliminary phylogenetic analyses performed on a pre-existing morphological dataset of modern and extinct cnidaria [14] consistently retrieved *Haootia* and *Mamsetia* as derived (crown-group) staurozoans (Supplementary Information S2).

7. Palaeobiological Implications

The depositional setting of the Ediacaran Conception Group in the Bonavista region of Newfoundland is that of a distal shelf or slope, which is dominated by turbidites and debrites [33–35]. The palaeoenvironment of deposition is compatible with the wide range of environments from which modern staurozoans are reported [111,112]. The palaeobiology and mode of life of many elements of the Ediacaran biota remain contentious, including: the harbouring of chemosymbionts [113], phagotrophy [113,114], osmotrophy [115,116], suspension feeding using specialized organs [23,51], whole-body suspension feeding [117], and/or gas exchange [117,118]. If we accept *Haootia* and *Mamsetia* as staurozoan cnidarians, they are likely to have had a primarily heterotrophic mode of life, either through the active predation on plankton or feeding on detrital material on the seafloor, by analogy with modern staurozoa [106,119–121].

The preserved muscles of both the arms and primary tentacles of *Haootia* are consistent with active locomotion and with our assertion that it did not, in fact, have a large basal anchoring disk, contra [29,46]. If *Haootia* and *Mamsetia* are accepted as stauromedusae, a small pedal disc was likely present at the end of the peduncle.

8. Phylogenetic Implications

It is commonly agreed that the Metazoa and Eumetazoa evolved in either the Cryogenian or the Tonian [122,123], an estimate that significantly predates their first appearance in the Ediacaran fossil record. The clearest evidence of an Ediacaran animal in the oldest known Avalon Assemblage biotas is in the form of trace fossils [27,63,64] and the medusozoan body fossil genera *Haootia, Mamsetia*, and *Auroralumina* [14,29]. Metazoan affinities have also been proposed for: *Thectardis* [124], the Arboreomorpha [125], and the Rangeomorpha [12].

Traditionally, the Porifera have been considered a monophyletic group, diverging from other animals as early as in the Cryogenian [126,127], but this has been challenged by some recent studies, which see Porifera as a paraphyletic group in the metazoan stem instead [114,128]. The paraphyly hypothesis indirectly supports Cavalier-Smith's [129] hypothesis that the eumetazoans evolved from a pre-sponge-grade ancestor. Some studies propose the Ctenophora as the earliest group of animals [130–133], with sponges evolving later with the other animal groups in the late Ediacaran. This second hypothesis has been proposed as evidence to explain the apparent absence of sponges from the rich Ediacaran fossil record [134], but see the *Protospongia*-like *Helicolocellus cantori* [135]. The presence of staurozoan cnidarians in the Ediacaran fossil record does, however, require the presence of one or both of their two possible sister groups, the Porifera and the Ctenophora, even though their phylogenetic positions relative to the Metazoa are still under debate [130,131,136]. The presence of the genera *Haootia* and *Mamsetia* in the Ediacaran thus has important implications for the understanding of early animal evolution and the calibration of molecular clocks, but also for the evolutionary timing and phylogeny of the Cnidaria [137].

Staurozoans are traditionally considered to be an early-diverging monophyletic group within the subphylum Medusozoa, an interpretation which has been largely confirmed by molecular data [69,138], though the phylogenetic relationships within the Medusozoa are still debated [49,139].

Since the Anthozoa is a sister group to the Medusozoa, anthozoans are expected in the Ediacaran, with the inferred divergence date for the two groups thought to be during the Tonian [140]. The absence of confirmed anthozoan fossils from the Ediacaran can be explained by the late evolution of skeletal structures and the simple external morphology of the Actinaria [64].

If staurozoans are accepted as primitive medusozoans, they would play an important role in the debate over whether a polyp or a medusa was the plesiomorphic trait of the Cnidaria (and therefore the Medusozoa). It has been suggested that the Medusozoa might have evolved from a polyp-like ancestor, with the absence of a medusoid stage in the Anthozoa being the ancestral state rather than the result of later evolutionary loss [141], however, the stauromedusae of *Haootia* and *Mamsetia* are large and have well-developed muscular structures in the arms.

9. Conclusions

Haootia quadriformis is one of the most iconic of Ediacaran taxa and the earliest example of a staurozoan cnidarian in the fossil record, to which we now add a second genus, *Mamsetia manunis*. The exceptional preservation of muscle tissues in both genera allows the recognition of a tetra-radial symmetry that supports their inclusion in the Cnidaria. This work rejects the hypothesis that *Haootia* was attached to the seafloor by a large holdfast disc, and instead identifies a staurozoa-like peduncle. The suggestion that *Haootia* had very highly developed marginal muscles unlike those of extant stauromedusae [106] was based on misinterpretation of the distribution of the longitudinal muscles associated with the arms of the holotype, and was rejected herein following the recognition of delicate marginal/coronal musculature similar to that of modern stauromedusae. The marginal branches of *Haootia quadriformis* were here reinterpreted as being analogous to the perradial primary tentacles that lie at the margin of the calyx of modern Staurozoa. Primary tentacles are, as yet, unknown in *Mamsetia manunis*.

Haootia quadriformis differs from modern Staurozoa in having eight arms and at least two perradial marginal tentacles associated with the coronal muscles of each side of the calyx, traits which could be considered plesiomorphic to the class. *Mamsetia manunis* has the conventional four arms of the Staurozoa, but with no perradial musculature or primary tentacles preserved. We therefore consider that *Haootia* and *Mamsetia* are the best-preserved candidates for being the earliest medusozoans, and that they were morphologically similar to modern Staurozoa. The presence of cnidarians in the Ediacaran implies that Ctenophora and Porifera had already evolved during the Neoproterozoic, even though they might have been excluded from the peculiar Ediacaran taphonomic windows. The former paratype of *Haootia* might also conceivably be an early developmental (cf. stauropolyp) stage of either *Mamsetia* or *Haootia*, but further consideration requires the identification of more small specimens in the field.

Supplementary Materials: The following supporting information can be downloaded at: https://www.mdpi.com/article/10.3390/life14091096/s1, Supplementary Information S1: locomotion of *Lucernaria* sp.; and Supplementary Information S2: cladistical analysis encompassing *Haootia* and *Mamsetia*.

Author Contributions: D.M. and G.P., methodology; formal analysis, D.M. and G.P.; investigation, D.M., G.P., D.P.-P., C.M., S.C.D., J.J.M., L.R.M., R.N., and R.S.T.; resources, D.M., R.S.T., and G.P.; data curation, G.P.; writing—original draft preparation, D.M.; writing—review and editing, D.M., G.P., D.P.-P., C.M., S.C.D., J.J.M., L.R.M., R.N., and R.S.T.; visualization, D.M. and R.N.; supervision, D.M.; project administration, D.M.; funding acquisition, D.M. All authors have read and agreed to the published version of the manuscript.

Funding: This research is supported by the NSERC Discovery Grant #RGPIN-2018-04880.

Data Availability Statement: Data are contained within the article and Supplementary Materials.

Acknowledgments: This work was supported by an NSERC DG grant to D. McIlroy. Bob Hooper is thanked for his biological mentorship and advice, and the Memorial University Ocean Sciences dive team are recognized for their work in collecting *Lucernaria* for comparative study. This work was completed under permit from the Province of Newfoundland and Labrador. A. Liu is thanked for his donation of the cast of the holotype, based on his silicone mold of the type specimen, and F. Dunn for interesting early discussions on the affinities of *Haootia*.

Conflicts of Interest: The authors declare no conflicts of interest.

References

1. McIlroy, D.; Horàk, J.M. 2006 Neoproterozoic: The late Precambrian terranes that formed Eastern Avalonia. In *The Geology of England and Wales*; Brenchley, P.R., Rawson, P.F., Eds.; Geological Society Publishing House: Bath, UK, 2006; pp. 9–25.
2. Nance, R.D.; Murphy, J.B.; Keppie, D.J. A Cordilleran model for the evolution of Avalonia. *Tectonophysics* **2002**, *352*, 11–31. [CrossRef]
3. Pollock, J.C.; Hibbard, J.P.; van Staal, C.R. A paleogeographical review of the peri-Gondwanan realm of the Appalachian Orogen. *Can. J. Earth Sci.* **2012**, *49*, 259–289. [CrossRef]
4. Ford, T.D. Precambrian fossils from Charnwood Forest. *Proc. Yorks. Geol. Soc.* **1958**, *31*, 211–217. [CrossRef]
5. Ford, T.D. The Pre-Cambrian fossils of Charnwood Forest. *Trans. Leicester. Lit. Phil. Soc.* **1963**, *57*, 57–62. [CrossRef]
6. Boynton, H.E.; Ford, T.D. Ediacaran fossils from the Precambrian (Charnian Supergroup) of Charnwood Forest, Leicestershire, England. *Mercian Geol.* **1995**, *13*, 156–182.
7. Brasier, M.D.; Antcliffe, J.B. Evolutionary relationships within the Avalonian Ediacara biota: New insights from laser analysis. *J. Geol. Soc.* **2009**, *166*, 363–384. [CrossRef]
8. Wilby, P.R.; Carney, J.N.; Howe, M.P.A. A rich Ediacaran assemblage from eastern Avalonia: Evidence of early widespread diversity in the deep ocean. *Geology* **2011**, *39*, 655–658. [CrossRef]
9. Wilby, P.R.; Kenchington, C.G.; Wilby, R.L. Role of low intensity environmental disturbance in structuring the earliest (Ediacaran) microbenthic tiered communities. *Palaeogeogr. Palaeoclimatol. Palaeoecol.* **2015**, *434*, 14–27. [CrossRef]
10. Kenchington, C.G.; Wilby, P.R. Rangeomorph classification schemes and intra-specific variation: Are all characters created equal? In *Earth System Evolution and Early Life: A Celebration of the Work of Martin Brasier*; Brasier, A.T., McIlroy, D., Eds.; Special Publications of the Geological Society: London, UK, 2017; Volume 448, pp. 221–251.
11. Kenchington, C.G.; Dunn, F.S.; Wilby, P.R. Modularity and overcompensatory growth in Ediacaran rangeomorphs demonstrate early adaptations for coping with environmental pressures. *Curr. Biol.* **2018**, *28*, 3330–3336.e2. [CrossRef]
12. Dunn, F.S.; Wilby, P.R.; Kenchington, C.G.; Grazhdankin, D.V.; Donoghue, P.C.J.; Liu, A.G. Anatomy of the Ediacaran rangeomorph *Charnia masoni*. *Pap. Palaeontol.* **2019**, *5*, 157–176. [CrossRef]
13. Dunn, F.S.; Liu, A.G.; Grazhdankin, D.V.; Vixseboxse, P.; Flannery-Sutherland, J.; Green, E.; Harris, S.; Wilby, P.R.; Donoghue, P.C.J. The developmental biology of *Charnia* and the eumetazoan affinity of the Ediacaran rangeomorphs. *Sci. Adv.* **2021**, *7*, eabe0291. [CrossRef] [PubMed]
14. Dunn, F.S.; Kenchington, C.G.; Parry, L.A.; Clark, J.W.; Kendall, R.S.; Wilby, P.R. A crown-group cnidarian from the Ediacaran of Charnwood Forest, UK. *Nat. Ecol. Evol.* **2022**, *6*, 1095–1104. [CrossRef] [PubMed]

15. Anderson, M.M.; Misra, S.B. Fossils found in the Pre-Cambrian Conception Group of south-eastern Newfoundland. *Nature* **1968**, *220*, 680–681. [CrossRef]
16. Anderson, M.M.; Conway Morris, S. A review, with descriptions of four unusual forms, of the soft-bodied fauna of the Conception and St. John's groups (Late Precambrian), Avalon Peninsula, Newfoundland. *Proc. Third N. Am. Paleontol. Conv.* **1982**, *1*, 1–8.
17. Seilacher, A. Vendobionta and Psammocorallia-lost constructions of Precambrian evolution. *J. Geol. Soc. Lond.* **1992**, *149*, 607–613. [CrossRef]
18. Seilacher, A. Biomat-related lifestyles in the Precambrian. *Palaios* **1999**, *14*, 86–89. [CrossRef]
19. Laflamme, M.; Narbonne, G.M.; Anderson, M.M. Morphometric analysis of the Ediacaran frond *Charniodiscus* from the Mistaken Point Formation, Newfoundland. *J. Palaeontol.* **2004**, *78*, 827–837. [CrossRef]
20. Laflamme, M.; Narbonne, G.M.; Greentree, C.; Anderson, M.M. Morphology and taphonomy of an Ediacaran frond: *Charnia* from the Avalon Peninsula of Newfoundland. *Geol. Soc. Lond. Spec. Publ.* **2007**, *286*, 237–257. [CrossRef]
21. Laflamme, M.; Flude, L.I.; Narbonne, G.M. Ecological tiering and the evolution of a stem: The oldest stemmed frond from the Ediacaran of Newfoundland. *J. Paleontol.* **2012**, *86*, 193–200. [CrossRef]
22. Narbonne, G.M. Modular construction in the Ediacaran biota. *Science* **2004**, *305*, 1141–1144. [CrossRef]
23. Narbonne, G.M. The Ediacara biota: Neoproterozoic origin of animals and their ecosystems. *Ann. Rev. Earth Planet. Sci.* **2005**, *33*, 421–442. [CrossRef]
24. Hofmann, H.J.; O'Brien, S.J.; King, A.F. Ediacaran biota on the Bonavista Peninsula, Newfoundland, Canada. *J. Paleontol.* **2008**, *82*, 1–36. [CrossRef]
25. Narbonne, G.M.; Laflamme, M.; Greentree, C.; Trusler, P. Reconstructing a lost world: Ediacaran rangeomorphs from Spaniard's Bay, Newfoundland. *J. Paleontol.* **2009**, *83*, 503–523. [CrossRef]
26. Brasier, M.D.; Antcliffe, J.B.; Liu, A.G. The architecture of Ediacaran fronds. *Palaeontology* **2012**, *55*, 1105–1124. [CrossRef]
27. Liu, A.G.; McIlroy, D.; Brasier, M.D. First evidence for locomotion in the Ediacara biota from the 565Ma Mistaken Point Formation, Newfoundland. *Geology* **2010**, *38*, 123–126. [CrossRef]
28. Liu, A.G.; McIlroy, D.; Antcliffe, J.B.; Brasier, M.D. Effaced preservation in the Ediacaran biota and its implications for the early macrofossil record. *Palaeontology* **2011**, *54*, 607–630. [CrossRef]
29. Liu, A.G.; Matthews, J.J.; Menon, L.R.; McIlroy, D.; Brasier, M.D. The arrangement of possible muscle fibres in the Ediacaran taxon *Haootia quadriformis*. *Proc. R. Soc. B* **2015**, *282*, 20142949. [CrossRef]
30. Liu, A.G.; Matthews, J.J.; McIlroy, D. The *Beothukis/Culmofrons* problem and its bearing on Ediacaran macrofossil taxonomy: Evidence from an exceptional new fossil locality. *Palaeontology* **2016**, *59*, 45–58. [CrossRef]
31. Brasier, M.D.; Liu, A.G.; Menon, L.R.; Matthews, J.J.; McIlroy, D.; Wacey, D. Explaining the exceptional preservation of Ediacaran rangeomorphs from Spaniard's Bay, Newfoundland: A hydraulic model. *Precambrian Res.* **2013**, *231*, 122–135. [CrossRef]
32. McIlroy, D.; Brasier, M.D.; Moseley, J.B. The Proterozoic–Cambrian transition within the 'Charnian Supergroup' of central England and the antiquity of the Ediacara fauna. *J. Geol. Soc. Lond.* **1998**, *155*, 401–411. [CrossRef]
33. O'Brien, S.J.; King, A.F. Late Neoproterozoic (Ediacaran) stratigraphy of the Avalon Zone sedimentary rocks, Bonavista Peninsula, Newfoundland. *Curr. Res. Nfld. Labrador Dep. Nat. Resour. Geol. Surv.* **2005**, *05-1*, 101–113.
34. Ichaso, A.A.; Dalrymple, R.W.; Narbonne, G.M. Paleoenvironmental and basin analysis of the late Neoproterozoic (Ediacaran) upper Conception and St. John's groups, west Conception Bay, Newfoundland. *Can. J. Earth Sci.* **2007**, *44*, 25–41. [CrossRef]
35. Gehling, J.G.; Narbonne, G.M.; Anderson, M.M. The first named Ediacaran body fossil *Aspidella terranovica*. *Palaeontology* **2003**, *43*, 427–456. [CrossRef]
36. Liu, A.G.; Kenchington, C.G.; Mitchell, E.G. Remarkable insights into the paleoecology of the Avalonian Ediacaran macrobiota. *Gondwana Res.* **2015**, *27*, 1355–1380. [CrossRef]
37. McIlroy, D.; Dufour, S.C.; Taylor, R.S.; Nicholls, R. The role of symbiosis in the first colonization of the seafloor by macrobiota: Insights from the oldest Ediacaran biota (Newfoundland, Canada). *Biosystems* **2021**, *205*, 104413. [CrossRef] [PubMed]
38. Narbonne, G.M.; Gehling, J.G. Life after snowball: The oldest complex Ediacaran fossils. *Geology* **2003**, *31*, 27–30. [CrossRef]
39. Gehling, J.G.; Narbonne, G.M. Spindle shaped Ediacara fossils from the Mistaken Point assemblage, Avalon Zone, Newfoundland. *Can. J. Earth Sci.* **2007**, *44*, 367–387. [CrossRef]
40. McKean, C.; Taylor, R.S.; McIlroy, D. New taphonomic and sedimentological insights into the preservation of high-relief Ediacaran fossils at Upper Island Cove, Newfoundland. *Lethaia* **2023**, *56*, 1–17. [CrossRef]
41. Liu, A.G.; Dunn, F.S. Filamentous connections between Ediacaran fronds. *Curr. Biol.* **2016**, *30*, 1322–1328. [CrossRef]
42. Liu, A.G.; Matthews, J.J.; Menon, L.R.; McIlroy, D.; Brasier, M.D. *Haootia quadriformis* n. gen., n. sp., interpreted as a muscular cnidarian impression from the Late Ediacaran period (approx. 560 Ma). *Proc. R. Soc. B* **2014**, *281*, 20141202. [CrossRef]
43. Pasinetti, G.; McIlroy, D. Palaeobiology and taphonomy of the rangeomorph *Culmofrons plumosa*. *Palaeontology* **2023**, *66*, e12671. [CrossRef]
44. Pu, J.P.; Bowring, S.A.; Ramezani, J.; Myrow, P.; Raub, T.D.; Landing, E.; Mills, A.; Hodgin, R.; Macdonald, F.A. Dodging snowballs: Geochronology of the Gaskiers glaciation and the first appearance of the Ediacaran biota. *Geology* **2016**, *44*, 955–958. [CrossRef]
45. Matthews, J.J.; Liu, A.G.; Yang, C.; McIlroy, D.; Levell, B.; Condon, D.J. A chronostratigraphic framework for the rise of the Ediacaran macrobiota: New constraints from Mistaken Point Ecological Reserve, Newfoundland. *Geol. Soc. Am. Bull.* **2020**, *133*, 612–624. [CrossRef]

46. Miranda, L.S.; Collins, A.G.; Marques, A.C. Is *Haootia quadriformis* related to extant Staurozoa (Cnidaria)? Evidence from the muscular system reconsidered. *Proc. R. Soc. Biol.* **2015**, *282*, 20142396. [CrossRef]
47. Van Iten, H.; Leme, J.M.; Pacheco, M.L.A.F.; Simões, M.G.; Fairchild, T.R.; Rodrigues, F.; Galante, D.; Boggiani, P.C.; Marques, A.C. 2016 Origin and Early Diversification of phylum Cnidaria: Key macrofossils from the Ediacaran System of North and South America. In *The Cnidaria, Past Present, and Future*; Goffredo, S., Dubinsky, Z., Eds.; Springer: Cham, Switzerland, 2006; pp. 31–40. [CrossRef]
48. Kayal, E.; Bentlange, B.; Pankey, M.S.; Ohdera, A.H.; Medina, M.; Planchetzki, D.C.; Collins, A.G.; Ryan, J.F. Phylogenomics provides a robust topology of the major cnidarian lineages and insights on the origins of key organismal traits. *BMC Evol. Biol.* **2018**, *18*, 68. [CrossRef]
49. Kayal, E.; Roure, B.; Philippe, H.; Collins, A.G.; Lavrov, D.V. Cnidarian phylogenetic relationships as revealed by mitogenomics. *BMC Evol. Biol.* **2013**, *13*, 5. [CrossRef] [PubMed]
50. Grazhdankin, D. Forbidden fruits in the Garden of Ediacara. *PalZ* **2016**, *90*, 649–657. [CrossRef]
51. Glaessner, M.F. New fossils from the base of the Cambrian in South Australia. *Trans. R. Soc. S. Aust.* **1958**, *81*, 185–188.
52. Mawson, D.; Sprigg, R.C. Subdivision of the Adelaide System. *Aust. J. Sci.* **1950**, *13*, 69–72.
53. Antcliffe, J.B.; Brasier, M.D. Charnia at 50: Developmental models for Ediacaran fronds. *Palaeontology* **2008**, *51*, 11–26. [CrossRef]
54. Jenkins, R.J.F. The enigmatic Ediacaran (late Precambrian) genus *Rangea* and related forms. *Paleobiology* **1985**, *11*, 336–355. [CrossRef]
55. Hoyal Cuthill, J.F.; Conway Morris, S. Fractal branching organizations of Ediacaran rangeomorph fronds reveal a lost Proterozoic body plan. *Proc. Nat. Acad. Sci. USA* **2014**, *111*, 13122–13126. [CrossRef] [PubMed]
56. Taylor, R.S.; Nicholls, R.; Neville, J.M.; McIlroy, D. Morphological variation in the rangeomorph organism *Fractofusus misrai* from the Ediacaran of Newfoundland, Canada. *Geol. Mag.* **2023**, *160*, 146–166. [CrossRef]
57. Young, G.A.; Hagadorn, J.W. 2010 The fossil record of cnidarian medusa. *Palaeoworld* **2010**, *19*, 212–221. [CrossRef]
58. Burzynski, G.; Narbonne, G.M. The discs of Avalon: Relating discoid fossils to frondose organisms in the Ediacaran of Newfoundland, Canada. *Palaeogeogr. Palaeoclimatol. Palaeoecol.* **2015**, *434*, 34–45. [CrossRef]
59. Hancy, A.D.; Antcliffe, J.B. Anoxia can increase the rate of decay for cnidarian tissue: Using *Actinia equina* to understand the early fossil record. *Geobiology* **2020**, *18*, 167–184. [CrossRef]
60. Menon, L.R.; McIlroy, D.; Liu, A.G.; Brasier, M.D. The dynamic influence of microbial mats on sediments: Fluid escape and pseudofossil formation in the Ediacaran Longmyndian Supergroup, UK. *J. Geol. Soc.* **2016**, *173*, 177–185. [CrossRef]
61. Menon, L.R.; McIlroy, D.; Brasier, M.D. *'Intrites' from the Ediacaran Longmyndian Supergroup, UK: A new form of microbially-induced sedimentary structure (MISS)*; Special Publications of the Geological Society: London, UK, 2017; Volume 448, pp. 271–283. [CrossRef]
62. Fedonkin, M.A. White Sea biota of Vendian (Precambrian non-skeletal fauna of the Russian Platform North). *Trans. Geol. Inst. Acad. Sci. USSR* **1981**, *342*, 90040–90049.
63. Liu, A.G.; McIlroy, D. Horizontal surface traces from the Fermeuse Formation, Ferryland (Newfoundland, Canada) and their place within the later Ediacaran ichnological revolution. *Geol. Assoc. Can. Misc. Publ.* **2015**, *9*, 141–156.
64. Menon, L.R.; McIlroy, D.; Brasier, M.D. Evidence for Cnidaria-like behavior in ca. 560Ma Ediacaran *Aspidella*. *Geology* **2013**, *41*, 895–898. [CrossRef]
65. Hahn, G.; Hahn, R.; Leonardos, O.H.; Pflug, H.D.; Walde, D.H.G. Körperlich erhaltene scyphozoen-reste aus dem Jungpräkambrium Braziliens. *Geol. Palaeontol.* **1982**, *16*, 1–18.
66. Chen, Z.; Zhou, C.; Xiao, S.; Wang, W.; Guan, C.; Hua, H.; Yuan, X. New Ediacaran fossils preserved in marine limestone and their ecological implications. *Sci. Rep.* **2014**, *4*, 4180. [CrossRef]
67. Ivantsov, A.Y.; Fedonkin, M.A. Conulariid-like fossil from the Vendian of Russia: A metazoan clade across the Proterozoic/Palaeozoic boundary. *Palaeontology* **2002**, *45*, 1219–1229. [CrossRef]
68. Van Iten, H.; Leme, J.M.; Rodrigues, S.C.; Simões, M.G.; Simões, M.G. Reinterpretation of a conulariid-like fossil from the Vendian of Russia. *Palaeontology* **2005**, *48*, 619–622. [CrossRef]
69. Van Iten, H.; Leme, J.M.; Simões, M.G.; Marques, A.C.; Collins, A.G. Reassessment of the phylogenetic affinities of conulariids (?Ediacaran-Triasssic) within the subphylum Medusozoa (Phylum Cnidaria). *J. Syst. Palaeontol.* **2006**, *4*, 109–118. [CrossRef]
70. Leme, J.M.; Van Iten, H.; Simões, M.G. A new conulariid (Cnidaria, Scyphozoa) from the terminal Ediacaran of Brazil. *Front. Earth Sci.* **2022**, *10*, 777746. [CrossRef]
71. Han, J.; Kubota, S.; Uchida, H.; Stanley, G.D.; Yao, X.; Shu, D.; Li, Y.; Yasui, K. Tiny sea anemone from the lower Cambrian of China. *PLoS ONE* **2010**, *5*, 313276. [CrossRef]
72. Conway Morris, S. Ediacaran-like fossils in Cambrian Burgess Shale-type faunas of North America. *Paleontology* **1993**, *36*, 593–635.
73. Matthews, J.J.; Liu, A.G.; McIlroy, D. Post-fossilization processes and their implications for understanding Ediacaran macrofossil assemblages. *Geol. Soc. Lond. Spec. Publ.* **2017**, *448*, 251–269. [CrossRef]
74. McKean, C.; Matthews, J.J.; Rideout, B.W.T.; Taylor, R.S.; McIlroy, D. Protecting Newfoundland's palaeontological geoheritage: Mitigating anthropic and environmental damage to the Ediacaran fossils of Upper Island Cove. *Geoheritage* **2024**, *16*, 45. [CrossRef]
75. McIlroy, D.; Hawco, J.; McKean, C.; Nicholls, R.; Pasinetti, G.; Taylor, R.S. Palaeobiology of the reclining rangeomorph *Beothukis* from the Ediacaran Mistaken Point Formation of southeastern Newfoundland. *Geol. Mag.* **2023**, *159*, 1160–1174. [CrossRef]

76. Mason, S.J.; Narbonne, G.M.; Dalrymple, R.W.; O'Brien, S.J. Paleoenvironmental analysis of Ediacaran strata in the Catalina Dome, Bonavista Peninsula, Newfoundland. *Can. J. Earth Sci.* **2013**, *50*, 197–212. [CrossRef]
77. Pérez-Pinedo, D.; McKean, C.; Taylor, R.S.; Nicholls, R.; McIlroy, D. *Charniodiscus* and *Arborea* are separate genera within the Arboreomorpha: Using the holotype of *C. concentricus* to resolve a taphonomic/taxonomic tangle. *Front. Earth Sci.* **2022**, *9*, 785929. [CrossRef]
78. Pérez-Pinedo, D.; Neville, J.M.; Pasinetti, G.; McKean, C.; Taylor, R.S.; McIlroy, D. Frond orientations with independent current indicators demonstrate the reclining rheotropic mode of life of several Ediacaran rangeomorph taxa. *Paleobiology* **2023**, *49*, 471–492. [CrossRef]
79. McIlroy, D.; Pérez-Pinedo, D.; Pasinetti, G.; McKean, C.; Taylor, R.S.; Hiscott, R.N. Rheotropic epifaunal growth, not felling by density currents, is responsible for many Ediacaran fossil orientations at Mistaken Point. *Front. Earth Sci.* **2022**, *10*, 849194. [CrossRef]
80. Seilacher, A. 1984 Late Precambrian and Early Cambrian Metazoa: Preservational or real extinctions? In *Patterns of Change in Earth Evolution*; Holland, H.D., Trendall, A.F., Eds.; Springer: Berlin, Germany, 1984; pp. 159–168.
81. Seilacher, A.; Reif, W.-E.; Westphal, F. Sedimentological, ecological and temporal patterns of fossil Lagerstätten. *Philos. Trans. R. Soc. B* **1985**, *311*, 5–24.
82. Cai, Y.; Schiffbauer, J.D.; Hong, H.; Xiao, S. Morphology and paleoecology of the late Edicaran tubular fossil *Conotubus hemiannulatus* from the Gaojiashan Lagerstätte of southern Shaanxi Province, South China. *Precambrian Res.* **2010**, *191*, 46–57. [CrossRef]
83. Brett, C.E.; Zambito, J.J.; Hunda, B.R.; Schlindler, E. Mid-Paleozoic trilobite lagerstatten: Models of diagenetically enhanced obrution deposits. *Palaios* **2012**, *27*, 326–345. [CrossRef]
84. Fraga, M.C.; Vega, C.S. How does rapid burial work? New insights from experiments with echinoderms. *Palaeontology* **2024**, *67*, e12698. [CrossRef]
85. Sierra, N.C.; Gould, D.A. The evolution of cnidarian stinging cells supports a Precambrian radiation of animal predators. *Evol. Dev.* **2024**, *26*, e12469. [CrossRef]
86. Brusca, R.C.; Brusca, G.J. *2003 Invertebrates;* Sinauer Associates: Dunderland, MA, USA, 2003; 936p.
87. Wade, M. Preservation of soft-bodied animals in Precambrian sandstones at Ediacara, South Australia. *Lethaia* **1968**, *1*, 238–267. [CrossRef]
88. Bobrovskiy, I.; Krasnova, A.; Ivantsov, A.; Luzhnaya (Serezhnikova), E.; Brocks, J.J. Simple sediment rheology explains the Ediacara biota preservation. *Nat. Ecol. Evol.* **2019**, *3*, 582–589. [CrossRef] [PubMed]
89. McIlroy, D.; Brasier, M.D.; Lang, A.S. Smothering of microbial mats by macrobiota: Implications for the Ediacara biota. *J. Geol. Soc.* **2009**, *166*, 1117–1121. [CrossRef]
90. Antcliffe, J.B.; Hancy, A.D.; Brasier, M.D. A new ecological model for the ~565Ma Ediacaran biota of Mistaken Point, Newfoundland. *Precambrian Res.* **2015**, *268*, 227–242. [CrossRef]
91. Westlake, H.E.; Page, L.R. Muscle and nerve net organization in stalked jellyfish (Medusozoa: Staurozoa). *J. Morphol.* **2017**, *278*, 29–49. [CrossRef]
92. Sansom, R.S.; Freedman, K.; Gabbott, S.E.; Aldridge, R.J.; Purnell, M.A. Taphonomy and affinity of an enigmatic Silurian vertebrate, *Jamoytius kerwoodi* White. *Palaeontology* **2010**, *53*, 1393–1409. [CrossRef]
93. Nanglu, K.; Lerosey-Aubril, R.; Weaver, J.C.; Ortega-Hernández, J. A mid Cambrian tunicate and the deep origin of the ascidiacean body plan. *Nat. Commun.* **2023**, *14*, 3832. [CrossRef]
94. Zhang, X.-G.; Smith, M.R.; Yang, J.; Hou, J.B. Onychophoran-like musculature in a phosphatized Cambrian lobopodan. *Biol. Lett.* **2016**, *12*, 201660492. [CrossRef]
95. Parry, L.; Smithwick, F.; Norden, K.K.; Saitta, E.T.; Lozano-Fernandez, J.; Tanner, A.R.; Caron, J.-B.; Edgecombe, G.D.; Briggs, D.E.G.; Vinther, J. Soft-bodied fossils are not simply rotten carcasses–toward a holistic understanding of exceptional fossil preservation. *BioEssays* **2018**, *40*, 1700167. [CrossRef]
96. Hyman, L.H. Observations and experiments on the physiology of medusae. *Biol. Bull.* **1940**, *79*, 282–292. [CrossRef]
97. Ling, S.W. Studies on Chinese stauromedusae. I. Stauromedusae from Tsingtao. *Amoy Mar. Biol. Bull.* **1937**, *3*, 1–35.
98. Hou, X.-G.; Stanley, G.D., Jr.; Zhao, J.; Ma, X.-Y. Cambrian anemones with preserved soft tissue from the Chengjiang biota, China. *Lethaia* **2005**, *38*, 193–203. [CrossRef]
99. Larson, R.J. A new Stauromedusa, *Kishinouya corbini* (Scyphozoaa, Stauromedusae) from the tropical western Atlantic. *Bull. Mar. Sci.* **1980**, *30*, 102–107.
100. Fu, D.; Tong, G.; Dai, T.; Liu, W.; Yang, Y.; Zhang, Y.; Cui, L.; Li, L.; Yun, H. The Quingjiang biota: A Burgess Shale-type fossil lagerstätte from the early Cambrian of South China. *Science* **2019**, *363*, 1338–1342. [CrossRef]
101. Moon, J.; Caron, J.-B.; Moysiuk, J. A macroscopic free-swimming medusa from the middle Cambrian Burgess Shale. *Proc. R. Soc. B* **2023**, *290*, 20222490. [CrossRef] [PubMed]
102. Raikova, E.V. Life cycle, cytology and morphology of *Polypodium hydriforme*, a coelenterate parasite of the eggs of acipenseriform fishes. *J. Parasitol.* **1994**, *80*, 1–22. [CrossRef]
103. Raikova, E.V. Cytomorphological peculiarities of *Polypodium hydriforme* (Cnidaria). *J. Mar. Biol. Assoc. UK* **2008**, *88*, 1695–1702. [CrossRef]

104. Raikova, E.V.; Ibragimov, A.Y.; Raikova, O.I. Muscular system of a peculiar parasitic cnidarian *Polypodium hydriforme*: A phalloidin fluorescence study. *Tissue Cell* **2007**, *39*, 79–87. [CrossRef]
105. Leclère, L.; Röttinger, E. Diversity of cnidarian muscles: Function, anatomy, development and regeneration. *Front. Cell Dev. Biol.* **2017**, *4*, 157. [CrossRef]
106. Gwilliam, G.F. Neuromuscular physiology of a sessile scyphozoan. *Biol. Bull.* **1960**, *119*, 454–473. [CrossRef]
107. Miranda, L.S.; Morandini, A.C.; Marques, A.C. Taxonomic review of *Haliclystus antacticus* Pfeffer, 1889 (Stauromedusae, Staurozoa, Cnidiaria) with remarks on the genus *Haliclystus* Clark, 1863. *Polar Biol.* **2009**, *32*, 1507–1519. [CrossRef]
108. Kikinger, R.; Salvini-Plawen, L. von, Development from polyp to stauromedusa in *Stylocoronella* (Cnidaria: Scyphozoa). *J. Mar. Biol. Assoc. UK* **1995**, *75*, 899–912. [CrossRef]
109. Wietrzykowski, W. Recherches sur le développement des lucernaires. *Arch. Zool. Exp. Gén.* **1912**, *5e Série 10*, 1–95.
110. Gershwin, L.A. Clonal and population variation in jellyfish symmetry. *J. Mar. Biol. Assoc. UK* **1999**, *79*, 993–1000. [CrossRef]
111. Thiel, H. 1966 The evolution of Scyphozoa—A review. In *The Cnidaria and Their Evolution*; Rees, W.J., Ed.; Academic Press: London, UK, 1966; pp. 77–117.
112. Lutz, R.A.; Collins, A.G.; Annis, E.R.; Reed, A.J.; Bennett, K.F.; Halanych, K.M.; Vrijenhoes, R.C. Stauromedusan populations inhabiting deep-sea hydrothermal vents along the southern East Pacific Rise. *Cah. Biol. Mar.* **2006**, *47*, 409–413.
113. Miranda, L.S.; Mills, C.E.; Hirano, Y.M.; Collins, A.G.; Marques, A.C. A review of the global diversity and natural history of stalked jellyfishes (Cnidaria, Staurozoa). *Mar. Biodiv.* **2018**, *48*, 1695–1714. [CrossRef]
114. Dufour, S.C.; McIlroy, D. Ediacaran pre-placozoan diploblasts in the Avalonian biota: The role of chemosynthesis in the evolution of early animal life. *Geol. Soc. Lond. Spec. Publ.* **2017**, *448*, 211–221. [CrossRef]
115. Sperling, E.A.; Vinther, J. A placozoan affinity for *Dickinsonia* and the evolution of later Proterozoic metazoan feeding modes. *Evol. Dev.* **2010**, *12*, 201–209. [CrossRef]
116. Laflamme, M.; Xiao, S.; Kowalewski, M. Osmotrophy in modular Ediacara organisms. *Proc. Nat. Acad. Sci. USA* **2009**, *106*, 14438–14443. [CrossRef]
117. Butterfield, N.J. Constructional and functional anatomy of Ediacaran rangeomorphs. *Geol. Mag.* **2022**, *159*, 1148–1159. [CrossRef]
118. Pérez-Pinedo, D.; Nicholls, R.; Neville, J.M.; McIlroy, D. Hydrodynamic insights into the paleobiology of the Ediacaran rangeomorph *Fractofusus misrai*. *iScience* **2024**, *24*, 110107. [CrossRef] [PubMed]
119. Darroch, S.A.F.; Guatarra, S.; Masaki, H.; Olaru, A.; Gibson, B.M.; Dunn, F.S.; Mitchell, E.G.; Racicot, R.A.; Burzynski, G.; Rahman, I.A. The rangeomorph *Pectinifrons abyssalis:* Hydrodynamic function at the dawn of animal life. *iScience* **2023**, *26*, 105989. [CrossRef] [PubMed]
120. Hirano, Y.M. Studies on stauromedusae (Coelenterata, Scyphozoa) in Hokkaido- Distribution and life history. *Benthos Res.* **1986**, *30*, 10–13. [CrossRef]
121. Larson, R.J.; Fautin, D.G. Stauromedusae of the genus *Manania* (=*Thaumatoscyphus*) (Cnidaria, Scyphozoa) in the Northeast Pacific, including descriptions of new species *Manania gwilliami* and *Manania handi*. *Can. J. Zool.* **1989**, *67*, 1543–1549. [CrossRef]
122. Zagal, C.J. Diet of the stauromedusa *Haliclystus auricula* from southern Chile. *J. Mar. Biol. Assoc. UK* **2004**, *84*, 337–340. [CrossRef]
123. dos Reis, M.; Thawornwattana, Y.; Angelis, K.; Telford, M.J.; Donoghue, P.C.J.; Yang, Z. Uncertainty in the timing of origin of animal and the limits of precision in molecular timescales. *Curr. Biol.* **2015**, *25*, 293902950. [CrossRef]
124. Cummings, J.A.; Liu, A.G.; Bengtson, S.; Donoghue, P.C.J. The origin of animals: Can molecular clocks and the fossil record be reconciled. *BioEssays* **2017**, *39*, 1600120.
125. Clapham, M.E.; Narbonne, G.M.; Gehling, J.G.; Greentree, C.; Anderson, M.M. *Thectardis avalonensis*: A new Ediacaran fossil from the Mistaken Point biota, Newfoundland. *J. Paleontol.* **2004**, *78*, 1031–1036. [CrossRef]
126. Dunn, F.S.; Liu, A.G.; Gehling, J.G. Anatomical and ontogenetic reassessment of the Ediacaran frond *Arborea arborea* and its placement within total group Eumetazoa. *Palaeontology* **2019**, *62*, 851–865. [CrossRef]
127. Wallace, M.W.; Hood, A.v.S.; Woon, E.M.S.; Hoffmann, K.-H.; Reed, C.P. Enigmatic chambered structures in Cryogenian reefs: The oldest sponge-grade organisms? *Precamb. Res.* **2014**, *225*, 109–123. [CrossRef]
128. Wörheide, G.; Dohrmann, M.; Erpenbeck, D.; Larroux, C.; Maldonado, M.; Voigt, O. Deep phylogeny and evolution of sponges (Phylum Porifera). *Adv. Mar. Biol.* **2012**, *2012*, 1–78. [CrossRef]
129. Sperling, E.A.; Pisani, D.; Peterson, K.J. Poriferan paraphyly and its implications for Precambrian palaeobiology. *Geol. Soc. Lond. Spec. Publ.* **2007**, *286*, 355–368. [CrossRef]
130. Cavalier-Smith, T. Origin of animal multicellularity: Precursors, causes, consequences—The choanoflagellate/sponge transition, neurogenesis and the Cambrian explosion. *Philos. Trans. R. Soc. B* **2017**, *372*, 20150476. [CrossRef]
131. Philippe, H.; Derelle, R.; Lopez, P.; Pick, K.; Borchiellini, C.; Boury-Esnault, N.; Vacelet, J.; Renard, E.; Houliston, E.; Quéinnec, E.; et al. Phylogenomics revives traditional views on deep animal relationships. *Curr. Biol.* **2009**, *19*, 706–712. [CrossRef]
132. Jákely, G.; Paps, J.; Nielsen, C. The phylogenetic position of ctenophores and the origin(s) of nervous systems. *EvoDevo* **2015**, *13*, 9135702. [CrossRef] [PubMed]
133. Whelan, N.V.; Halanych, K.M. Available data do not rule out Ctenophora as the sister group to all other Metazoa. *Nat. Commun.* **2023**, *14*, 711. [CrossRef]
134. Redmond, R.K.; McLysaght, A. Reply to: Available data do not rule out Ctenophora as the sister group to all other Metazoa. *Nat. Commun.* **2023**, *14*, 710. [CrossRef]
135. Botting, J.P.; Muir, L.A. Early sponge evolution: A review and phylogenetic framework. *Palaeoworld* **2018**, *27*, 1–29. [CrossRef]

136. Wang, X.; Liu, A.G.; Chen, Z.; Wu, C.; Liu, Y.; Wan, B.; Pang, K.; Zhou, C.; Youan, X.; Xiao, S. A late-Ediacaran crown-group sponge animal. *Nature* **2024**, *630*, 905–911. [CrossRef]
137. Pisani, D.; Pett, W.; Dohrmann, M.; Feuda, R.; Rota-Stabelli, O.; Philippe, H. Genomic data do not support comb-jellies as the sister group to all other animals. *Proc. Natl. Acad. Sci. USA* **2015**, *112*, 15402–15407. [CrossRef]
138. Pisani, D.; Liu, A.G. Animal evolution: Only rocks can set the clock. *Curr. Biol.* **2015**, *25*, 1079–1081. [CrossRef]
139. Miranda, L.S.; Hirano, Y.M.; Mills, C.E.; Falconer, A.; Fenwick, D.; Marques, A.C.; Collins, A.G. Systematics of stalked jellyfishes (Cnidaria: Staurozoa). *PeerJ* **2016**, *4*, e1951. [CrossRef]
140. Zapata, F.; Goetz, F.E.; Smith, A.S.; Howison, M.; Siebert, S.; Church, S.H. Phylogenic analyses support traditional relationships within the Cnidaria. *PLoS ONE* **2015**, *10*, e0139068. [CrossRef]
141. Macfadden, C.; Quattrini, A.M.; Brugler, M.R.; Cowman, P.F.; Dueñas, L.F.; Kitahara, M.V.; Paz-García, D.A.; Reimer, J.D.; Rodríguez, E. Phylogenomics, origin and diversification of anthozoans (Phylum Cnidaria). *Syst. Biol.* **2021**, *70*, 635–647. [CrossRef]

life

MDPI

Review

A Worldwide Annotated Checklist of Fossil (Devonian–Cretaceous) Species of the Clam Shrimp Genus *Euestheria* (Branchiopoda: Diplostraca: Spinicaudata)

Shaohua Fang [1,2] and Huanyu Liao [1,*]

[1] Institute of Palaeontology, Yunnan Key Laboratory of Earth System Science, Yunnan Key Laboratory for Palaeobiology, MEC International Joint Laboratory for Palaeobiology and Palaeoenvironment, Yunnan University, Kunming 650500, China

[2] School of Earth Science, Yunnan University, Kunming 650500, China

* Correspondence: hyliao@ynu.edu.cn

Abstract: Clam shrimps are one of the most common and representative invertebrates in continental strata and are endowed with important biostratigraphic and paleoecological values. The genus *Euestheria* is one of the most common clam shrimp taxa that has been recorded in the latest Paleozoic and the Mesozoic around the world. A list of all the species assigned to *Euestheria*, recorded from the Devonian to the Cretaceous across all seven continents as of September 2024, is provided. The list may serve as a valuable resource, potentially useful for the biostratigraphic division and long-range correlations of continental strata. In addition, the taxonomic issues of the genus are briefly introduced and examined herein. The study aims to provide a simple lead-in for all the researchers who have an interest in clam shrimp and the genus *Euestheria*.

Keywords: fossil record; arthropod; conchostracan; palaeogeographic distribution; stratigraphic distribution

1. Introduction

Clam shrimps, or "conchostracans," are a paraphyletic group of bivalved crustaceans (Figure 1a–d) known in fossil records from the Paleozoic to recent times. Modern clam shrimps are usually seen in small lakes, ponds, swamps, temporary pools, and paddy fields with still water. Fossil clam shrimps are one of the most common and representative invertebrates in continental strata and are endowed with important biostratigraphic and paleoecological values [1–8]. The global fossil record indicates that the evolutionary history of these crustaceans can be at least traced back to the Early Devonian (Emsian) [8–15]. The genus *Euestheria* Depéret & Mazeran, 1912 is one of the most common taxa that has been recorded in the latest Paleozoic and the Mesozoic around the world (Figure 2) [2,4,5,11,16–21]. However, there is currently no comprehensive reference for the taxonomy and geographical and stratigraphic distributions of this "well-known" taxon. This study provides the first primary checklist of the global fossil record of the genus *Euestheria* based on an extensive literature survey, with a peculiar emphasis on their geographical and stratigraphic distributions. The checklist may serve as a valuable resource, potentially useful for the biostratigraphic division and long-range correlations of continental strata. Moreover, the taxonomic issues are briefly examined and discussed herein. All these works aim to provide a simple lead-in for all the researchers who have an interest in fossil clam shrimps and the genus *Euestheria*.

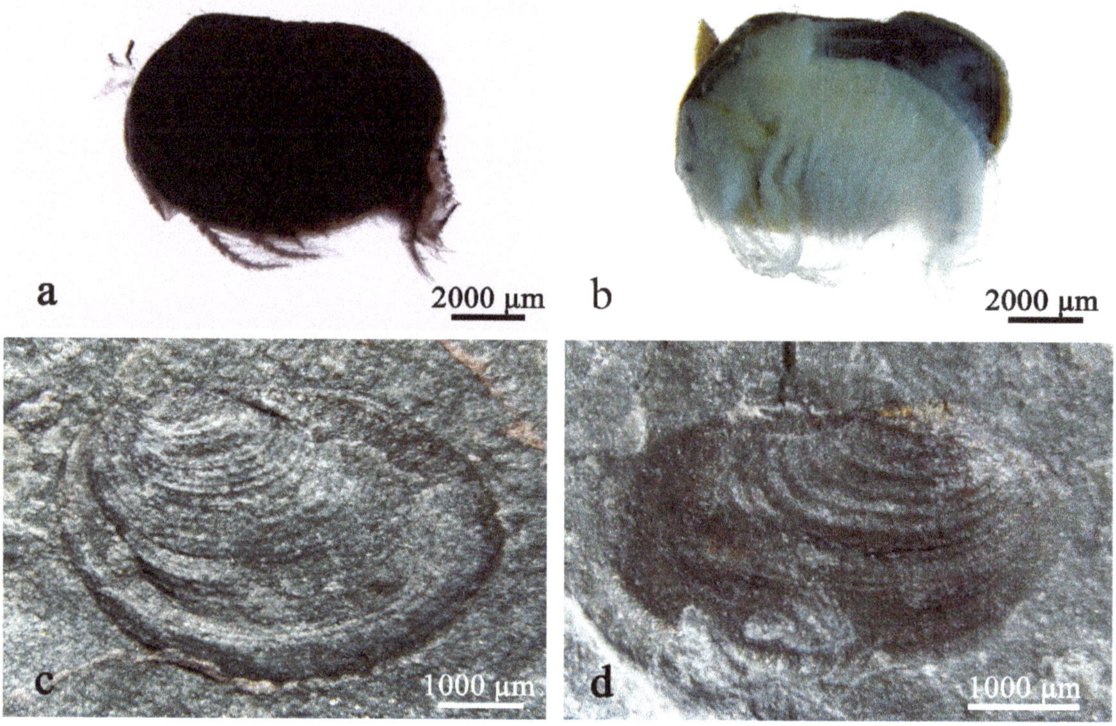

Figure 1. Modern and fossil specimens of clam shrimp: (**a**). a modern spinicaudatan with a whole carapace, from Tibet, China; (**b**). a modern spinicaudatan with the left valve of the carapace removed, showing "soft parts" of its body, from Tibet, China; (**c**). fossil specimen of *Euestheria* sp. from the Triassic Badong Formation in Hubei, China, showing the left valve of the carapace (housed at the Palaeobotanical Collections of the Institute of Palaeontology, Yunnan University, China, under catalog number YNUPB20001); (**d**). fossil specimen of *Euestheria* sp. from the Triassic Badong Formation in Hubei, China, showing the right valve of the carapace (housed at the Palaeobotanical Collections of the Institute of Palaeontology, Yunnan University, China, under catalog number YNUPB20002).

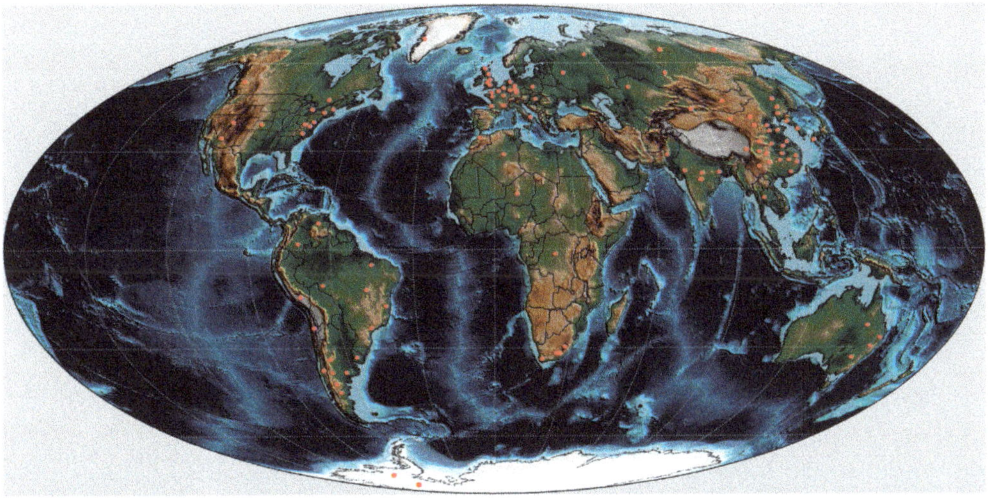

Figure 2. Global geographical distribution of all the *Euestheria* species records reviewed in the present paper shown on a map of present-day continental configurations (the map is based on Scotese [22]).

2. Result

2.1. Checklist

Our study shows that a total of 249 species (217 named species including 10 new combinations and 32 names in open nomenclature (23 names with "conf." and 9 with "aff.")) have been assigned to *Euestheria*, formally recorded from the Devonian to the Cretaceous of all seven continents (Table 1). Two pie charts were made to show the percentages of species numbers geographically and chronologically (Figure 3). It is worth mentioning that our statistics and checklist may have several issues due to factors like the age of the publications, language barriers, and challenges in data collection caused by difficulties in accessing the literature. If readers notice any mistakes or missing information, we encourage them to reach out to us.

Table 1. Worldwide distributional and stratigraphical checklist of *Euestheria* species listed in chronological order based on the date of their description.

Species	Horizon, Age, and Locality	Reference
Euestheria minuta (Zieten, 1833)	Early–Middle Triassic, Europe (Germany and France); Lacoste borehole, Late Triassic, Gard, France; Vouhenans, borehole, Late Triassic, Haute-Saône, Vosges, France; Conliège borehole, Triassic, Jura, France; Triassic, Bex, Switzerland; Triassic, western High-Atlas, Morocco; Early Jurassic, Indre, France; Cape Biot Fm., Late Triassic, Greenland; Xujiahe Fm., Late Triassic, Sichuan, China; Ganhaizi Fm., Xiangyun Fm., Maichujing Fm., Manghuai Fm., Late Triassic, Yunnan, China; Feixianguan Fm., Yongningzhen Fm., Early Triassic, Guizhou, China; Yanchang Gr., Middle–Late Triassic, Gansu, China; Jiapeila Fm., Bagong Fm., Late Triassic, Tibet, China; Guanyintan Fm., Late Triassic, Hunan, China; Blizzard Heights Fm., Early Jurassic, Antarctica; Nanyinger Gr., Late Triassic, Gansu, China; Dingjiayao Fm., Middle Triassic, Gansu, China; Potrerillos Fm., Santa Clara de Abajo Fm., Middle–Late Triassic, Argentina; Wolfville Fm., Middle Triassic, Canada; Devonian, New York State, USA; Genkou Gr., Late Triassic, Guangdong, China; Tuckahoe Fm., Middle–Late Triassic, USA; Haojiagou Fm., Late Triassic, Xinjiang, China; Panchet Fm., Pali Fm., Late Permian–Early Triassic, Raniganj Coalfield, India; Santa Maria Fm., Middle Triassic, Brazil	Zieten, 1833; Defretin, 1953, 1969; Cuvelier et al., 2015; Katoo, 1971; Chen, 1974; Zhang et al., 1976; Luo et al., 2018; Shen, 1981; Gu & Cheng, 1981; Tasch, 1987; Cai, 1990, 1993; Gallego, 1992, 1999, 1999; Knox & Gordon, 1999; Zhang & Lin, 2000; Kozur & Weems, 2007, 2010; Sun et al., 2010; Ghosh, 2011; Jenisch et al., 2017. [2,4,5,14,18,19,23–38]

Table 1. *Cont.*

Species	Horizon, Age, and Locality	Reference
Euestheria cf. *minuta*	Tongchuan Fm., Middle Triassic, Shaanxi, China; Xingkoushi Fm., Early Jurassic, Hebei, China; Blina Shale Fm., Early Triassic, Australia; Cacheuta Fm., Middle–Late Triassic, Argentina; Potrerillos Fm., Late Triassic, Argentina	Wang & Liu, 1980; Shen & Chang, 1993; Tasch, 1987; Gallego, 1999, 1999; Tassi et al., 2015. [19,32,33,39–41]
Euestheria aff. *minuta*	Santa Maria Fm., Middle Triassic, Brazil	Katoo, 1971. [26]
Euestheria albertii (Voltz, 1835)	Moenkopi Fm., Middle Triassic, Arizona, USA	Voltz, 1835; Lucas & Schoch, 2002; Kozur & Weems, 2010. [4,42,43]
Euestheria albertii mahlerselli Kozur & Lepper	Solling Fm, Early Triassic, Franconia, Bavaria, Germany	Kozur & Mock, 1993. [44]
Euestheria vatus (Lea, 1856)	Triassic, USA	Lea, 1856; Spamer, 1989. [45,46]
Euestheria brodieana (Jones, 1862)	Triassic, Gloucestershire, Warwickshire, England; Yanchang Gr., Late Triassic, Shaanxi, China; Yanchang Fm., Late Triassic, Shaanxi, China; Midland Fm., Late Triassic, England; Waterfall Fm., Early Jurassic, France, Germany; Late Triassic, USA	Jones, 1862; Wang & Liu, 1980; Pang, 1993; Kozur & Weems, 2007, 2011. [36,39,47–49]
Euestheria middendorfii (Jones, 1862)	Jurassic, Turgay Plateau, Kazakhstan	Jones, 1862. [47]
Euestheria mangaliensis (Jones, 1862)	Triassic, India; Early Triassic, Wardha Basin, India; Early Triassic, Nágpur. India; Argentina; Sunjiagou Fm., Permian, Shaanxi, China; Kamthi Fm., Early Triassic, Berwickshire, England	Jones, 1862; Tasch, 1987; Gallego, 1992; Liu & He, 2000; Ghosh, 2011. [5,19,31,47,50]
Euestheria kotnhensis (Jones, 1862)	Early Jurassic, India; Middle Triassic, Kota, India	Jones, 1862; Kobayashi, 1954. [16,47]
Euestheria forbesi (Jones, 1862)	Mesozoic, South America; Late Triassic, Mendoza, Argentina; Ischichuca Fm., Middle–Late Triassic, La Rioja, Argentina	Jones, 1862; Tasch, 1987; Gallego et al., 2001. [19,47,51]
Euestheria cf. *forbesi*	Santa Maria Fm., Middle Triassic, Brazil	Gallego, 1996. [52]
Euestheria murchisoniae (Jones, 1862)	Late Jurassic, Scotland	Jones, 1862; Kobayashi, 1954. [16,47]
Euestheria orientalis Eichwald, 1868	Jurassic, Nertschinsk, Russia	Eichwald, 1868; Kobayashi, 1954. [16,53]
Euestheria dawsoni (Jones, 1870)	Canso Gr., Carboniferous, New Hampshire, USA; Cheverie Fm., Canada; Anzin Fm., Flines Fm., Carboniferous, Pas-de-Calais, France	Jones,1870; Copeland, 1957; Bell, 1960; Cuvelier et al., 2015. [14,54–56]
Euestheria rimosa (Goldenberg, 1877)	Steinkohlen Fm., Permian, Saarland, Germany	Goldenberg, 1877. [57]
Euestheria limbata (Goldenberg, 1877)	Steinkohlen Fm., Permian, Saarland, Germany	Goldenberg, 1877. [57]
Euestheria greyi (Jones, 1878)	Late Permian, Cape Town, South Africa	Jones, 1878; Tasch, 1987. [19,58]
Euestheria karpinskiana (Jones, 1883)	Triassic, Siberia, Russia	Jones, 1883; Kobayashi, 1954. [16,59]
Euestheria laxitecta (Jones, 1890)	Triassic, Gipskeuper, Germany; Late Triassic, England; Late Triassic, Krasiejów, Poland.	Jones, 1890; Olempska, 2004. [60,61]
Euestheria ? *laitexta*	Xujiahe Fm., Late Triassic, Chongqing, China	Chen, 1974; Zhang et al., 1976. [2,18]

Table 1. *Cont.*

Species	Horizon, Age, and Locality	Reference
Euestheria cebnnensis (Grand & Eury, 1890)	Lower Stephanian, Carboniferous, La Mure Basin, France	Grand, 1890. [62]
Euestheria anomala (Jones, 1901)	Early Cretaceous, Heidelberg, Swellendam, South Africa	Jones, 1901; Tasch, 1987. [19,63]
Euestheria simoni Pruvost, 1911	Bruay Fm., Westphalian C (now late Moscovian), Carboniferous, Nord, France; Rabat, Morocco; South Wales, England	Pruvost, 1911; Cuvelier et al., 2015; Dix & Trueman, 1928; Tasch, 1987. [14,19,64–66]
Euestheria kawasakii Ozawa & Watanabe, 1923	Mesozoic, Korea; Xiangyun Fm., Late Triassic, Yunnan, China; Amison Fm., Late Triassic, South Korea	Ozawa & Watanabe, 1923; Zhang et al., 1976; Kim & Lee, 2015. [2,67,68]
Euestheria elongata Tchernychev, 1926	Permian, Russia	Tchernychev, 1926. [69]
Euestheria ? elongata	Baitianba Fm., Early Jurassic, Sichuan, China	Zhang et al., 1976. [2]
Euestheria novocastrensis Mitchell, 1927	Newcastle Coal Measures, Dirty Coal seam, Late Permian, New South Wales, Australia	Mitchell, 1927; Tasch, 1987. [19,70]
Euestheria ipsviciensis Mitchell, 1927	Late Triassic, Denmark Hill, Ipswich, Queensland, Australia	Mitchell, 1927; Tasch, 1987. [19,70]
Euestheria belmontensis Mitchell, 1927	Late Permian, Dirty Coal seam, New South Wales, Australia	Mitchell, 1927; Tasch, 1987. [19,70]
Euestheria trigonellaris Mitchell, 1927	Late Permian, Dirty Coal seam, New South Wales, Australia	Mitchell, 1927; Kobayashi, 1954; Tasch, 1987. [16,19,70]
Euestheria nyuicastrensis Mitchell, 1927	Late Permian, New South Wales, Australia	Mitchell, 1927; Kobayashi, 1954. [16,70]
Euestheria obliqua Mitchell, 1927	Late Permian, New South Wales, Australia	Mitchell, 1927; Kobayashi, 1954. [16,70]
Euestheria dahurica Chernyshev, 1930	Early Cretaceous, Russia	Chernyshev, 1930. [71]
Euestheria cf. *dahurica*	Early Cretaceous, Africa	Gallego et al., 2020. [6]
Euestheria nengkiangensis Chi, 1931	Nenjiang Fm., Early Cretaceous, Heilongjiang, China	Ji, 1931. [72]
Euestheria sinkiangensis Chi, 1931	Late Jurassic, Turpan, Xinjiang, China; Sangonghe Fm., Early Jurassic, Xinjiang, China	Ji, 1931; Zhang et al., 1976; Wei, 1984. [2,72,73]
Pseudestheria diensti Poschmann, 2024 (=*Euestheria diensti* Gross, 1934)	Devonian, Germany	Gross,1934; Poschmann et al., 2024 [3,74]
Euestheria gutta (Lyutkevich, 1937)	Calvrd Fm., Early Triassic, Germanic Basin, Germany; Jaworznia Fm., Late Permian, Holy Cross Mountains, Poland; Sunjiagou Fm., Late Permian, Shaanxi, China; Kayitou Fm., Early Triassic, Guizhou, China; Vokhma Fm., Obnora Fm., Early Triassic, Russia; Charkabozhskaya Fm., Early Triassic, Myla Rive, Komi Republic, Russia	Lvutkeyich, 1937; Kozur & Seidel, 1983; Ptaszyński & Niedźwiedzki, 2004; Chu et al., 2013; Chu et al., 2018; Scholze et al., 2020; Miao et al., 2021; Scholze et al., 2015, 2017; Vladimirovna et al., 2020. [15,75–83]

Table 1. *Cont.*

Species	Horizon, Age, and Locality	Reference
Euestheria aff. *gutta*	Kayitou Fm., Early Triassic, Guizhou, China	Scholze et al., 2020. [80]
Euestheria aequale (Lyutkevich, 1937)	Permian, Russia	Lvutkeyich, 1937. [75]
Euestheria evenkiensis (Lyutkevich, 1937)	Late Permian, Late Tatarian (now Changhsingian), Central Siberia, Russia	Lvutkeyich, 1937; Cuvelier et al., 2015. [14,75]
Euestheria hausmanni Schmidt, 1938	Cumnock Fm., Lockatong Fm., New Oxford Fm., Late Triassic, Coburg Sandstein, Germany	Schmidt, 1928; Kozur & Weems, 2007. [36,84]
Euestheria stockmansi Maillieux, 1939	Early Devonian, Belgium; Devonian, Vincy, France; Early Triassic, Australia	Maillieux, 1939; Raymond, 1946; Kobayashi, 1954; Defretin, 1950; Novojilov, 1958. [10,11,16,85,86]
Euestheria subsmoni Deleau, 1945	Early Carboniferous, Algeria	Deleau, 1945. [87]
Euestheria exsecta Novojilov, 1946	Triassic, Russia; Volpriehausen Fm. and Detfurth Fm., Early Triassic, Germanic Basin, Germany; Hardegsen Fm., Early Triassic, Siberia, Russia; Csopak Marl Fm., Early Triassic, Hungary	Novojilov, 1946; Lucas & Schoch, 2002; Kozur & Weems, 2010. [1,4,43]
Euestheria multinstila Novojilov, 1946	Permian, Russia	Novojilov, 1946; Kobayashi, 1954. [1,16]
Euestheria belli Raymond, 1946	Cheverie Fm., Richmond County, Canada	Raymond, 1946; Bell, 1960. [11,56]
Euestheria autunensis Raymod, 1946	Permian, France	Raymond, 1946. [11]
Euestheria anchietai Teixeira, 1947	Late Permian, Africa	Teixeira, 1947. [88]
Euestheria striolatissima Rusconi, 1948	Middle Triassic, Argentina	Rusconi, 1948. [89]
Euestheria lerichei Marliere, 1950	Late Cretaceous, Africa	Mouta & Marlière, 1950. [90]
Euestheria passaui Marliere, 1950	Jurassic, Congo	Mouta & Marlière, 1950; Egoroff & Lombard, 1961. [90,91]
Euestheria ricourr Defretin, 1950	Triassic, France	Defretin, 1950. [12]
Euestheria mendesi Almerida, 1950	São Paulo, Brazil	Almerida, 1950. [92]
Euestheria khinganensis Kobayashi, 1951	Jurassic, Jilin, China	Kobayashi, 1951; Zhang et al., 1976. [2,93]
Euestheria rampoensis Kobayashi, 1951	Gabisan Fm., South Korea; Amisan Fm., Late Triassic, South Korea	Kobayashi, 1951; Kim & Lee, 2015. [68,93]
Euestheria aff. *rampoensis*	Middle Jurassic, Hebei, China	Wang et al., 1984. [94]
Euestheria shimamurai Kobayashi, 1951	Late Triassic, North Korea; Xiangyun Fm., Late Triassic, Yunnan, China; Amison Fm., Late Triassic, South Korea	Kobayashi, 1951; Kobayashi, 1954; Zhang et al., 1976; Kim & Lee, 2015. [2,16,68,93]
Euestheria cf. *shimamurai*	Nanyinger Gr., Late Triassic, Gansu, China	Kobayashi, 1951; Zhang et al., 1976. [2,93]

Table 1. *Cont.*

Species	Horizon, Age, and Locality	Reference
Euestheria atsuensis Kobayashi, 1951	Triassic, Japan	Kobayashi, 1951. [93]
Euestheria ? cf. *atsuensis*	Xujiahe Fm., Late Triassic, Sichuan, China	Zhang et al., 1976. [2]
Congestheriella olsoni Bock, 1954 (=*Euestheria olsoni* Bock, 1953)	La Quinta Fm., Late Triassic or Early Jurassic, Andean Cordillera, Venezuela	Bock, 1953; Gallego et al., 2010. [95,96]
Euestheria princetonensis Bock, 1953	Lockatong Fm., Late Triassic, New Jersey, USA	Bock, 1953. [95]
Euestheria winterpockensis Bock, 1953	Late Triassic, Virginia, USA; Grabfeld Fm., Tuckahoe Fm., and Falling Creek Fm., Late Triassic, Germany; Passaic Fm. and Bull Run Fm., Late Triassic, New Jersey, USA; Bocas Fm. and Montebel Fm., Late Triassic, Colombia	Bock, 1953; Kozur & Weems, 2007; Weems & Lucas, 2015; Alarcón et al., 2024. [36,95,97,98]
Euestheria eifelensis Novojilov, 1953 (=*Palaeolimnadiopsis eifelensis* Raymond, 1946)	Early Devonian, Germany; Devonian, Russia	Raymond, 1946; Novojilov, 1953. [11,99]
Euestheria chop rensis Novojilov, 1953	Devonian, Russia	Novojilov, 1953. [99]
Euestheria sainsllalidensis Novojilov, 1953	Mongolia	Novojilov, 1953. [99]
Euestheria aff. *sainsllalidensis*	Jurassic, Russia	Stepanov, 1966. [100]
Euestheria kidoi Kobayashi, 1954	Late Triassic, Korea	Kobayashi, 1954. [16]
Euestheria kusumi Kobayashi, 1954	Late Triassic, Korea	Kobayashi, 1954. [16]
Euestheria langei Mendes, 1954	Permian, Brazil	Mendes, 1954. [101]
Euestheria janovi Novojilov, 1955	Devonian, Russia	Novojilov, 1955. [102]
Euestheria consummata Novojilov, 1955	Devonian, Russia	Novojilov, 1955. [102]
Euestheria ltakassica Novojilov, 1955	Devonian, Russia	Novojilov, 1955. [103]
Euestheria leonardii Tasch, 1958 (=*Euestheria harveyi wellingtoni* Tasch, 1956)	Permian, Kansas, USA	Tasch, 1956, 1958. [104,105]
Euestheria leonardii Tasch, 1958 (=*Euestheria harveyi harveyi* Tasch, 1956)	Permian, Kansas, USA	Tasch, 1956, 1958. [104,105]

Table 1. *Cont.*

Species	Horizon, Age, and Locality	Reference
Euestheria leonardii Tasch, 1956	Permian, Kansas, USA	Tasch, 1956, 1958. [104,105]
Euestheria lamberti Defretin, 1956	Agades, Africa	Defretin et al., 1956; Defretin, 1958. [106,107]
Euestheria marginata Defretin, 1956	Agades, Africa	Defretin et al., 1956; Defretin, 1958. [106,107]
Euestheria azambujai Pinto, 1956	Santa Maria Fm., Late Triassic, Rio Grande do Sul, Brazil	Pinto, 1956. [108]
Euestheria raymondi Copeland, 1957	Canso Gr., Early Mississippian (Carboniferous), Canada	Copeland, 1957. [55]
Euestheria jakutica Kashirtsev, 1957	Early Triassic, Russia	Kashirtsev, 1957. [109]
Euestheria kaschirzwi Kashirtsev, 1957	Early Triassic, Russia	Kashirtsev, 1957. [109]
Euestheria lirella Bell, 1960	Cheverie Fm., Horton Gr., Canada	Bell, 1960. [56]
Euestheria cf. *lirella*	Albert Fm., New Brunswick, Canada	Greiner, 1974. [110]
Euestheria kokurensis Kusumi, 1960	Cretaceous, Kokura, Japan	Kusumi, 1960. [111]
Euestheria imamurai Kusumi, 1960	Cretaceous, Kokura, Japan	Kusumi, 1960. [111]
Euestheria taniiformis Zaspelova, 1961	Sangonghe Fm., Early Jurassic, Xingjiang, China; Wennan Fm., Early Jurassic, Shandong, China.	Zaspelova, 1961; Lu, 1995; Chen, 1982. [112–114]
Euestheria cf. *taniiformis*	Xingkoushi Fm., Early Jurassic, Hebei, China	Shen & Chang, 1993. [40]
Euestheria tenuiformis Zaspelova 1961	Sangonghe Fm., Early Jurassic, Xinjiang, China	Zaspelova, 1961; Chen, 1995. [112,115]
Euestheria spitzbergensis Zaspelova 1961	Late Triassic, Russia	Zaspelova, 1961. [112]
Euestheria ? taniiformis	Baitianba Fm., Shiwandashan Gr., Early Jurassic, Sichuan, Guangxi, China	Zhang et al., 1976. [2]
Euestheria aff. *taniiformis*	Badaowan Fm., Early Jurassic, Xinjiang, China	Chen, 2003. [116]
Euestheria franconica (Reible, 1962)	Meißner Fm., Middle Triassic, USA	Reible, 1962; Sell, 2018. [117,118]
Euestheria cf. *franconica*	Erfurt Fm., Middle Triassic, USA	Sell, 2018. [118]
Euestheria elliptica Molin, 1965	Permian, Russia; Ermaying Fm., Middle Triassic, Shaanxi, China	Molin, 1965; Wu, 1991. [119,120]
Euestheria osvanjensis Molin, 1965	Early Triassic, Russia	Molin, 1965. [119]
Euestheria udorica Molin, 1966	Early Triassic, Russia	Molin, 1966; Lipatova & Lopato, 2000. [121,122]
Euestheria sambaensis Simone, 1967	Cretaceous, Congo	Defretin, 1967. [123]
Euestheria sambaensis Defretin & Lefranc, 1967	Early Cretaceous, Africa	Defretin, 1967. [123]

Table 1. *Cont.*

Species	Horizon, Age, and Locality	Reference
Euestheria bourozi Defretin & Lefranc, 1970	Bruay Fm., Westphalian (now late Moscovian), Carboniferous, Pas-de-Calais, France	Cuvelier et al., 2015. [14]
Euestheria volkheimeri Tasch, 1970	Canadon Asfalto Fm., Jurassic, Patagonia, South America	Tasch, 1970. [124]
Euestheria syntchaense Novojilov, 1970	Late Triassic, Russia	Novojilov, 1970; Chunikhin, 2009. [125,126]
Euestheria spltzbergensis Zaspelova, 1972	Late Triassic, Russia	Zaspelova et al., 1972. [127]
Euestheria tumida Zaspelova, 1972	Late Triassic, Russia	Zaspelova et al., 1972. [127]
Euestherie ? luchangensis Chen, 1974	Baiguowan Fm., Late Triassic, Sichuan, China	Chen, 1974; Zhang et al., 1976. [2,18]
Euestheria yimengensis Chen, 1974	Baiguowan Fm., Late Triassic, Sichuan, China	Chen, 1974; Zhang et al., 1976. [2,18]
Euestheria contracta Chen, 1974	Baiguowan Fm., Late Triassic, Sichuan, China	Chen, 1974; Zhang et al., 1976. [2,18]
Euestheria weiyuanensis Chen, 1974	Xujiahe Fm., Late Triassic, Chongqing, China	Chen, 1974; Zhang et al., 1976. [2,18]
Euestheria dazuensis Chen, 1974	Xujiahe Fm., Late Triassic, Chongqing, China; Baiguowan Fm., Late Triassic, Sichuan, China; Ganhaizi Fm. and Xiangyun Fm., Late Triassic, Yunnan, China; Manghuai Fm., Middle Triassic, Yunnan, China	Chen, 1974; Luo et al., 2018. [18,27]
Euestheria yipinglangensis Chen, 1974	Xujiahe Fm., Late Triassic, Chongqing, China; Baiguowan Fm., Late Triassic, Sichuan, China; Ganhaizi Fm. and Xiangyun Fm., Late Triassic, Yunnan, China; Yanchang Gr., Late Triassic, Ningxia, China; Bagong Fm., Late Triassic, Tibet, China; Dingjiayao Fm., Middle Triassic, Gansu, China; Genkou Gr., Late Triassic, Guangdong, China; Manghuai Fm., Middle Triassic, Yunnan, China	Chen, 1974; Zhang et al., 1976; Chen & Shen, 1981; Cai, 1990; Zhang & Lin, 2000; Luo et al., 2018. [2,18,27,29,35,128]
Euestheria cf. *yipinglangensis*	Xingshikou Fm., Early Jurassic, Hebei, China	Shen & Chang, 1993. [40]
Euestheria hechuanensis Chen, 1974	Xujiahe Fm., Late Triassic, Chongqing, China	Chen, 1974; Zhang et al., 1976. [2,18]
Euestheria ? tianquanensis Chen, 1974	Xujiahe Fm., Late Triassic, Sichuan, China	Chen, 1974; Zhang et al., 1976. [2,18]
Euestheria buravasi Kobayashi, 1975	Nam Pha Fm. and Huai Hin Lat Fm., Late Triassic, Khorat Plateau, Thailand; Cumnock Fm. and Catharpin Creek Fm., Late Triassic, USA; Sanford Fm., Late Triassic, USA	Kobayashi, 1975; Kozur & Weems, 2007, 2010; Chitnarin et al., 2022. [4,36,129,130]
Euestheria mansuyi Kobayashi, 1975	Nam Pha Fm., Late Triassic, Nam Phrom Dam, Thailand	Kobayashi, 1975; Chonglakmani et al., 1990. [129,131]
Euestheria thailandica Kobayashi, 1975	Nam Pha Fm., Late Triassic, Nam Phrom Dam, Thailand	Kobayashi, 1975. [129]
Euestheria bunopasi (=*Cyclestherioides bunopasi* Kobayashi, 1975)	Nam Pha Fm., Late Triassic, Nam Phrom Dam, Thailand	Kobayashi, 1975, 1984; Burrett, 2021. [129,132,133]

Table 1. *Cont.*

Species	Horizon, Age, and Locality	Reference
Euestheria rhombica Chen, 1976	Yanchang Gr., Late Triassic, Gansu, China; Genkou Gr., Late Triassic, Guangdong, China	Zhang et al., 1976; Zhang & Lin, 2000. [2,35]
Euestheria mupangensis Chen, 1976	Xiangyun Fm. and Maichujing Fm., Late Triassic, Yunnan, China; Bagong Fm. and Jiapeila Fm., Tibet, China; Wenbinshan Fm., Late Triassic, Fujian, China	Zhang et al., 1976; Lv et al., 2016; Chen & Shen, 1981; Liang, 1998. [2,128,134,135]
Euestheria obliqua Chen, 1976	Xiangyun Fm., Late Triassic, Yunnan, China; Dirty Coal Seam, Late Permian, New South Wales, Australia; Genkou Gr., Late Triassic, Guangdong, China	Zhang et al., 1976; Tasch, 1987; Zhang & Lin, 2000. [2,19,35]
Euestheria angusta Chen, 1976	Xiangyun Fm., Late Triassic, Yunnan, China; Bagong Fm., Late Triassic, Tibet, China	Zhang et al., 1976; Chen & Shen, 1981. [2,128]
Euestheria cf. *angusta*	Xingshikou Fm., Early Jurassic, Hebei, China	Shen & Chang, 1993. [40]
Euestheria lata Chen, 1976	Ganhaizi Fm., Late Triassic, Yunnan, China; Bagong Fm., Late Triassic, Tibet, China; Dudley seam, Dirty Coal seam, Permian, Niger, Australia	Zhang et al., 1976; Chen & Shen, 1981; Tasch, 1987. [2,19,128]
Euestheria xiangyunensis Chen, 1976	Xiangyun Fm. and Maichujin Fm., Late Triassic, Yunnan, China	Zhang et al., 1976; Lv et al., 2016. [2,134]
Euestheria yanjingxiensis Shen, 1976	Xujiahe Fm., Late Triassic, Chongqing, China; Kayitou Fm., Early Triassic, Guizhou, China; Zhangping Fm., Middle Jurassic, Fujian, China	Zhang et al., 1976; Chu et al., 2013; Cao, 1989. [2,78,136]
Euestheria langdaiensis Shen, 1976	Feixianguan Fm., Early Triassic, Guizhou, China	Zhang et al., 1976. [2]
Euestheria langdaiensis magna	Bernburg Fm., Early Triassic, Germany	Kozur, 1980. [137]
Euestheria orbicula Shen, 1976	Yongningzhen Fm., Kayitou Fm., Early Triassic, Guizhou, China	Zhang et al., 1976; Chu et al., 2013. [2,78]
Euestheria leidaiyanensis Shen, 1976	Yongningzhen Fm., Early Triassic, Guizhou, China	Zhang et al., 1976. [2]
Euestheria hubeiensis Shen, 1976	Badong Fm., Middle Triassic, Hubei, China	Zhang et al., 1976. [2]
Euestheria cf. *hubeiensis*	Xujiashan Fm., Middle Triassic, Jiangsu, China	Min et al., 1981. [138]
Euestheria lepida Shen, 1976	Badong Fm., Middle Triassic, Hubei, China	Zhang et al., 1976. [2]
Euestheria dactylis Shen, 1976	Badong Fm., Middle Triassic, Hubei, China; Kayitou Fm., Early Triassic, Guizhou, China	Zhang et al., 1976; Chu et al., 2013. [2,78]
Euestheria multilinearis (Shen, 1976)	Badong Fm., Middle Triassic, Hubei, China	Zhang et al., 1976. [2]
Euestheria shizibaoensis Shen, 1976	Badong Fm., Middle Triassic, Hubei, China; Kayitou Fm., Early Triassic, Guizhou, China	Zhang et al., 1976; Chu et al., 2013. [2,78]
Euestheria cf. *shizibaoensis*	Xujiashan Fm., Middle Triassic, Jiangsu, China	Min et al., 1981. [138]
Euestheria sparsa Shen, 1976	Badong Fm., Middle Triassic, Hubei, China	Zhang et al., 1976. [2]
Triglypta shandanensis Liao, 2017 (=*Euestheria shandanensis* Chen, 1976)	Longfengshan Gr., Early–Middle Jurassic, Gansu, China; Wangjiashan Fm., Middle Jurassic, Gansu, China; Lishan Fm., Early Jurassic, China; Dongyingfang Fm., Middle Jurassic, Liaoning, China; Wenbinshan Fm., Late Triassic, Fujian, China; Xiaoping Fm., Late Triassic, Guangdong, China; Sangonghe Fm., Badaowan Fm., Early Jurassic, Xinjiang, China	Zhang et al., 1976; Li et al., 1982; Cao, 1989; Li, 1990; Liang, 1998; Zhang & Lin, 2000; Shen, 2003. [2,35,136,139–141]

Table 1. *Cont.*

Species	Horizon, Age, and Locality	Reference
Euestheria aff. *shandanensis*	Wennan Fm., Early Jurassic, Shandong, China	Chen, 1982. [114]
Euestheria cf. *shandanensis*	Xingshikou Fm. and Laohugou Fm., Early Jurassic, Hebei, Liaoning, China; Sangonghe Fm., Early Jurassic, Xinjiang, China	Shen & Chang, 1993; Fu, 1998. [40,142]
Euestheria ziliujingensis Chen, 1976	Haifanggou Fm., Middle Jurassic, Liaoning, China; Shaximiao Fm., Middle Jurassic, Guizhou, China; Xiashaximiao Fm., Middle Jurassic, Sichuan, Chongqing, China; Luoao Fm., Middle Jurassic, Jiangxi, China; Jiulongshan Fm., Middle Jurassic, Hebei, China; Zhangjiakou Fm., Late Jurassic, Liaoning, China; Qingtujin Gr., Middle Jurassic, China; Wanbao Fm., Middle Jurassic, Inner Mongolia, China; Longfengshan Fm., Middle Jurassic, Gansu, China; Xintiangou Fm., Middle Jurassic, Guizhou, China	Zhang et al., 1976; Feng, 1978; Lu, 1995; Huang et al., 1998; Shen et al., 2003; Shen, 2004; Bao et al., 2011; Zhang et al., 2012; Zhang et al., 2013; Zhang et al., 2016. [2,113,143–150]
Euestheria aff. *ziliujingensis*	Xiaodonggou Fm., Middle Jurassic, Liaoning, Jilin, China; Anning Fm., Tuodian Fm., Late Jurassic, Yunnan, China	Fan, 1996; Cheng et al., 2004. [151,152]
Triglypt haifanggouensis Liao, 2017 (=*Euestheria haifanggouensis* Chen, 1976)	Haifanggou Fm., Middle Jurassic, Liaoning, China; Zhanghe Fm., Middle Jurassic, Yunan, China; Wangjiashan Fm., Middle Jurassic, Gansu, China; Shaximiao Fm., Middle Jurassic, Guizhou, China; Dongyingfang Fm., Middle Jurassic, Liaoning, China; Luoao Fm., Middle Jurassic, Jiangxi China; Jiulongshan Fm., Middle Jurassic, Inner Mongolia, China; Wanbao Fm., Middle Jurassic, Inner Mongolia, China; Zhangjiakou Fm., Early Jurassic, Inner Mongolia, China	Zhang et al., 1976; Shen et al., 2003; Liao et al., 2017; Feng, 1978; Li, 1990; Huang et al., 1998; Bao et al., 2011; Zhang et al., 2013. [2,139,143–145,147,149,153]
Euestheria aff. *haifanggouensis*	Xiaodonggou Fm., Middle Jurassic, Liaoning, Jilin, China; Xintiangou Fm., Middle Jurassic, Guizhou, China	Fan, 1996; Zhang et al., 2016. [150,151]
Euestheria cf. *haifanggouensis*	Hadataolegai Fm., Middle Jurassic, Jilin, China	Tan et al., 2019. [154]
Euestheria rotunda Zhang, 1976	Shaximiao Fm., Middle Jurassic, Sichuan, China; Qingtujin Gr., Middle Jurassic, China	Zhang et al., 1976; Zhang et al., 2016; Shen, 2004. [2,146,150]
Euestheria ? subquadrata Chen, 1976	Wangjiashan Fm., Middle Jurassic, Gansu, China	Zhang et al., 1976. [2]
Euestheria changtanensis Shen,1976	Ziliujing Fm., Middle Jurassic, Sichuan, China	Zhang et al., 1976. [2]
Euestheria elegans Shen, 1976	Guizhou Gr., Middle Jurassic, Hubei, China	Zhang et al., 1976. [2]
Euestheria cf. *elegans*	Sangonghe Fm., Early Jurassic, Xinjiang, China	Fu, 1998. [143]
Euestheria orientalis Shen, 1976	Guizhou Gr., Middle Jurassic, Hubei, China	Zhang et al., 1976. [2]
Euestheria jingyuanensis Chen, 1976	Wangjiashan Fm., Middle Jurassic, Gansu, China; Toutunhe Fm., Middle Jurassic, Xinjiang, China; Hadataolegai Fm., Middle Jurassic, Jilin, China; Jiulongshan Fm., Middle Jurassic, Inner Mongolia, China; Shaximiao Fm., Middle Jurassic, Guizhou, Sichuan, China; Tuodian Fm., Late Jurassic, Yunnan, China; Xinhe Fm., Middle Jurassic, Gansu, China	Zhang et al., 1976; Feng, 1978; Lu, 1995; Shen et al., 2003; Cheng et al., 2004; Peng et al., 2016; Gao et al., 2017; Tan et al., 2019. [2,113,140,143,152,154–156]
Euestheria cf. *jingyuanensis*	Qingtujin Gr., Middle Jurassic, China	Shen, 2004. [146]
Euestheria fabiformis Chen, 1976	Wangjiashan Fm., Middle Jurassic, Gansu, China; Xinhe Fm., Middle Jurassic, Gansu, China	Zhang et al., 1976; Peng et al., 2016. [2,156]
Euestheria aff. *fabiformis*	Qingtujin Gr., Middle Jurassic, China	Shen, 2004. [146]

Table 1. *Cont.*

Species	Horizon, Age, and Locality	Reference
Euestheria yanjiawanensis Chen, 1976	Xiashaximiao Fm., Middle Jurassic, Sichuan, China; Wangjiashan Fm., Middle Jurassic, Gansu, China; Anning Fm., Late Jurassic, Yunnan, China; Qingtujin Gr., Middle Jurassic, China; Zhangping Fm., Middle Jurassic, Fujian, China	Zhang et al., 1976; Liang, 1998; Cheng et al., 2004; Shen, 2004; Gao et al., 2017. [2,135,147,153,156]
Euestheria ? cf. *yanjiawanensis*	Laohugou Fm., Early Jurassic, Liaoning, China	Li et al., 1982; Shen & Chang, 1993. [40,141]
Euestheria complanata Chen, 1976	Guangyuan Gr., Middle Jurassic, Sichuan, China; Zhangping Fm., Middle Jurassic, Fujian, China; Dongyingfang Fm., Middle Jurassic, Liaoning, China; Qingtujin Gr., Middle Jurassic, Inner Mongolia, China; Longfengshan Fm., Middle Jurassic, Gansu, China; Shaximiao Fm., Middle Jurassic, Sichuan, China	Zhang et al., 1976; Cao et al., 1989; Liang, 1998; Li, 1990; Shen, 2004; Zhang et al., 2013; Gao et al., 2017. [2,135,136,139,146,149,156]
Euestheria ? cf. *complanata*	Anning Fm., Late Jurassic, Yunnan, China	Li et al., 1982; Cheng et al., 2004. [141,152]
Euestheria xiazhuangensis Chen, 1976	Zhanghe Fm., Middle Jurassic, Yunnan, China	Zhang et al., 1976. [2]
Euestheria manzhuangensis Chen, 1976	Huakaizuo Fm., Middle Jurassic, Yunnan, China; Qiakemake Fm., Middle Jurassic, Xinjiang, China	Zhang et al., 1976; Liu, 1990. [2,157]
Euestheria cf. *manzhuangensis*	Anning Fm., Late Jurassic, Yunnan, China	Cheng et al., 2004. [152]
Euestheria exilis Chen, 1976	Huakaizuo Fm., Middle Jurassic, Yunnan, China	Zhang et al., 1976. [2]
Euestheria ? *yangbiensis* Chen, 1976	Huakaizuo Fm., Middle Jurassic, Yunnan, China	Zhang et al., 1976. [2]
Euestheria cf. *yangbiensis*	Tuoman Fm., Late Jurassic, Yunnan, China	Cheng et al., 2004. [152]
Euestheria ? *batangensis* Chen, 1976	Late Triassic, Sichuan, China	Zhang et al., 1976. [2]
Euestheria ? *favosa* Chen, 1976	Dalazi Fm., Early Cretaceous, Jilin, China	Zhang et al., 1976. [2]
Euestheria ? *ambiqua* Zhang & Chen, 1976	Nenjiang Fm., Early Cretaceous, Heilongjiang, China	Zhang et al., 1976. [2]
Euestheria datongensis Zhang, 1976	Yungang Fm., Middle Jurassic, Shanxi, China	Zhang et al., 1976. [2]
Euestheria ? *nanchuanensis* Chen, 1976	Shangshaximiao Fm., Middle Jurassic, Sichuan, China	Zhang et al., 1976; Gao et al., 2017. [2,156]
Euestheria fuyuanensis Chen,1976	Kayitou Fm., Early Triassic, Guizhou, Yunnan, China	Zhang et al., 1976; Scholze et al., 2020. [2,80]
Euestheria wuzaoensis Chen & shen, 1978	Wuzao Fm., Late Triassic, Zhejiang, China	Chen, 1978. [158]
Euestheria trinanguliformis Duan, 1978	Badong Fm., Middle Triassic, Sichuan, China	Duan, 1978. [159]
Euestheria nanxiensis Duan, 1978	Xujiahe Fm., Late Triassic, Sichuan, China	Duan, 1978. [159]

Table 1. *Cont.*

Species	Horizon, Age, and Locality	Reference
Euestheria dongyuemiaoensis Duan, 1978	Ziliujing Fm., Middle Jurassic, Sichuan, China	Duan, 1978. [159]
Euestheria zizhongensis Duan, 1978	Ziliujing Fm., Middle Jurassic, Sichuan, China	Duan, 1978. [159]
Euestheria jiangyouensis Duan, 1978	Xiashaximiao Fm., Middle Jurassic, Sichuan, China	Duan, 1978. [159]
Euestheria bijieensis Feng, 1978	Shaximiao Fm., Middle Jurassic, Guizhou, China	Feng, 1978. [143]
Euestheria ? xiangshuiensis Feng, 1978	Shaximiao Fm., Middle Jurassic, Guizhou, China	Feng, 1978. [143]
Euestheria ? oertlii Kozur, 1980	Bernburg Fm., Early Triassic, Germany	Kozur, 1980. [137]
Euestheria tongchuangensis Liu, 1980	Tongchuan Fm., Middle Triassic, Shaanxi, China	Wang & Liu, 1980. [39]
Euestheria gibba Liu, 1980	Tongchuan Fm., Middle Triassic, Shaanxi, China	Wang & Liu, 1980. [39]
Euestheria huanglongensis Liu, 1980	Tongchuan Fm., Middle Triassic, Shaanxi, China	Wang & Liu, 1980. [39]
Euestheria celeta Liu, 1980	Tongchuan Fm., Middle Triassic, Shaanxi, China	Wang & Liu, 1980. [39]
Euestheria jinsuoguanensis Liu, 1980	Tongchuan Fm., Middle Triassic, Shaanxi, China	Wang & Liu, 1980. [39]
Euestheria hanchengensis Liu, 1980	TongchuanFm., Middle Triassic, Shaanxi, China	Wang & Liu, 1980. [39]
Euestheria deforma Liu, 1980	Yanchang Gr., Late Triassic, Shaanxi, China	Wang & Liu, 1980. [39]
Euestheria shensiensis Liu, 1980	Yanchang Gr., Late Triassic, Shaanxi, China	Wang & Liu, 1980. [39]
Euestheria changhangouensis Wu, 1980	Changhangou Fm., Middle Jurassic, Inner Mongolia, China	Wu, 1980. [160]
Euestheria cf. *changhangouensis*	Mawa Fm., Middle Jurassic, Henan, China	Hu, 1991. [161]
Euestheria shiguaiziensis Wu, 1980	Changhangou Fm., Middle Jurassic, Inner Mongolia, China	Wu, 1980. [160]
Euestheria cf. *shiguaiziensis*	Mawa Fm., Middle Jurassic, Henan, China	Hu, 1991. [161]
Euestheria jiangdaensis Chen & Shen, 1981	Jiapeila Fm., Late Triassic, Tibet, China	Chen & Shen, 1981. [128]
Euestheria deyiensis Chen & Shen, 1981	Jiapeila Fm., Late Triassic, Tibet, China	Chen & Shen, 1981. [128]

Table 1. *Cont.*

Species	Horizon, Age, and Locality	Reference
Euestheria waxianensis Chen & Shen, 1981	Shangshaximiao Fm., Middle Jurassic, Sichuan, China	Chen & Shen, 1981. [128]
Euestheria shandongensis Chen, 1982	Wennan Fm., Early Jurassic, Shandong, China.	Chen, 1982. [114]
Euestheria rongxianensis Xu, 1982	Xiashaximiao Fm., Middle Jurassic, Sichuan, China	Xu, 1982. [162]
Euestheria truempyi Shen, 2002 (=*Magniestheria truempyi* Kozur & Seidel, 1983)	Early Triassic, Germany	Kozur & Seidel, 1983; Shen, 2002. [76,163]
Euestheria jimsarensis Wei, 1984	Huangshanjie Fm., Late Triassic, Xinjiang, China	Wei, 1984. [73]
Euestheria stockleyi Tasch, 1984	Late Triassic, Africa	Tasch, 1984. [164]
Euestheria ? chaohuensis Shen, 1985	Hanshan Fm., Middle Jurassic, Anhui, China	Lu et al., 1985. [165]
Euestheria ? pengzhuanensis Shen, 1985	Hanshan Fm., Middle Jurassic, Anhui, China	Lu et al., 1985. [165]
Euestheria taschi Vallati, 1986	Cañadón Asfalto Fm., Chubut, Argentina	Vallati, 1986; Gallego, 2010. [96,166]
Euestheria thabaningensis Tasch, 1987	Late Triassic, Africa	Tasch, 1987. [19]
Euestheria triassibrevis Tasch, 1987	Maji ya Chumvi Fm., Early Triassic, Africa	Tasch, 1987. [19]
Triglypta luanpingensis Liao, 2017 (=*Euestheria luanpingensis* Shen & Niu, 1987)	Jiulongshan Fm., Middle Jurassic, Hebei, Inner Mongolia, China	Zhang et al., 1987; Shen et al., 2003; Liao et al., 2017. [145,153,166]
Euestheria dakongensis Cao, 1987	Lishan Fm., Early Jurassic, Fujian, China; Xiaoping Fm., Late Triassic, Guangdong, China	Cao, 1986; Liang, 1998; Zhang & Lin, 2000. [35,135,167]
Euestheria multicostata Geyer, 1987	Sanford Fm., Amstadt Fm., Late Triassic, Thuringia, Germany	Kozur & Weems, 2007. [36]
Euestheria lashlyensis Tasch, 1987	Middle–Late Devonian, Lashly Mountains, Antarctica	Tasch, 1987. [19]
Euestheria ritchiei Tasch, 1987	Late Devonian, South Victoria Land, Antarctica	Tasch, 1987. [19]
Euestheria juravariabalis Tasch, 1987	Early Jurassic, Mauger Nunatak, Antarctica	Tasch, 1987. [19]
Euestheria bearmorensis Tasch, 1987	Early Jurassic, BH Station, Blizzard Heights, Antarctica	Tasch, 1987. [19]
Euestheria ellioti Tasch, 1987	Early Jurassic, BH Station, Blizzard Heights, Antarctica	Tasch, 1987. [19]

Table 1. *Cont.*

Species	Horizon, Age, and Locality	Reference
Euestheria juracircularis Tasch, 1987	Early Jurassic, BH Station, Blizzard Heights, Antarctica	Tasch, 1987. [19]
Euestheria formavariabalis Tasch, 1987	Early Jurassic, BH Station, Blizzard Heights, Antarctica	Tasch, 1987. [19]
Euestheria ichthystromatos Tasch, 1987	Early Jurassic, Station 0, Storm Peak, Antarctica	Tasch, 1987. [19]
Euestheria crustapatulis Tasch, 1987	Early Jurassic, Storm Peak, Antarctica	Tasch, 1987. [19]
Euestheria rhadinis Tasch, 1987	Early Jurassic, Storm Peak, Antarctica	Tasch, 1987. [19]
Euestheria transantarctensis Tasch, 1987	Early Jurassic, Storm Peak, Antarctica	Tasch, 1987. [19]
Euestheria castaneus Tasch, 1987	Early Jurassic, Storm Peak, Antarctica	Tasch, 1987. [19]
Euestheria talenti Tasch, 1987	Devonian, Victoria, Australia	Tasch, 1987. [19]
Euestheria abaetensis Tasch, 1987 (=*Platyestheria* abaetensis Cardoso, 1971)	Quirico Fm., Early Cretaceous, Sanfranciscana Basin, Brazil	Tasch, 1987; Carvalho, 1993; Cardoso, 1971. [19,168,169]
Euestheria crustapatulus Tasch, 1987	Ferrar Gr., Storm Peak, Antarctica	Tasch, 1987; Carvalho, 1993. [19,168]
Euestheria thabaningensis Tasch, 1987	Late Triassic, Lesotho	Tasch, 1987. [19]
Euestheria triassibrevis Tasch, 1987	Narrabeen Gr., Early Triassic, Port Hacking, Australia	Tasch, 1987. [19]
Euestheria basbatiliensis Tasch, 1987	Panchet Fm., Early Triassic, Raniganj Coal Field, India	Tasch, 1987. [19]
Euestheria raniganjis Tasch, 1987	Panchet Fm., Early Triassic, Raniganj Coal Field, India	Tasch, 1987. [19]
Euestheria dualis Tasch, 1987	Panchet Fm., Early Triassic, Raniganj Coal Field, India	Tasch, 1987. [19]
Euestheria crustabundis Tasch, 1987	Kotá Fm., Early Jurassic, India	Tasch, 1987. [19]
Euestheria lefranci Tasch, 1987	Late Cretaceous, Algeria, Africa	Tasch, 1987. [19]
Euestheria aricensis (Jones) Tasch, 1987	Permian or Triassic, Arica, Arequipa, Peru	Tasch, 1987. [19]
Euestheria cf. *aricensis*	Permian, Chile	Tasch, 1987. [19]
Euestheria nitida Wu, 1991	Ermaying Fm., Middle Triassic, Shaanxi, China	Wu, 1991. [120]

Table 1. *Cont.*

Species	Horizon, Age, and Locality	Reference
Euestheria brevielliptica Wu, 1991	Ermaying Fm., Middle Triassic, Shaanxi, China	Wu, 1991. [120]
Punctatestheria trotternishensis Zhang, 2017 (=*Euestheria trotternishensis* Chen & Hudson, 1991)	Lealt Shale Fm., Middle Jurassic, Skye, Scotland	Chen & Hudson, 1991; Zhang et al., 2017. [170,171]
Euestheria falconeri Gallego, 1993	Yaguarı Fm., Late Permian, Uruguay	Gallego et al., 1993. [172]
Euestheria rocablanquensis Gallego, 1994	Roca Blanca Fm., Early–Middle Jurassic, Santa Cruz, Argentina	Gallego, 1994. [173]
Euestheria covacevichi Gallego, 1994	Peine Fm., Late Permian, Antofagasta, Chile	Gallego & Breitkreuz, 1994. [174]
Euestheria santamariensis Gallego, 1996	Santa Maria Fm., Middle Triassic, Brazil	Gallego, 1996. [52]
Euestheria pricei Gallego, 1999	Castellanos Fm., Early Cretaceous, Uruguay	Gallego et al., 1999. [175]
Euestheria feysi Vannier, 2003	Carboniferous, Saône-et-Loire, France	Vannier et al., 2003. [176]
Euestheria martinsnetoi Gallego, 2004	Ria Mendzo Fm., Middle Triassic, Mendoza, Argentina	Gallego et al., 2004. [177]
Euestheria urengoic Chunikhin, 2004	Late Triassic, Western Siberia, Russia	Chunikhin, 2004. [178]
Euestheria duqinshanensis Shen, 2004	Qingtujin Gr., Middle Jurassic, Inner Mongolia, China	Shen, 2004. [146]
Euestheria acampestira Bishop, 2010	Early–Middle Triassic, Aranbanga Volcanics Gr., Queensland, Australia	Bishop, 2010. [179]
Euestheria sanrensis Ghosh, 2011	Pali Fm., Permian–Early Jurassic, Jharkhand, India	Ghosh, 2011. [5]
Euestheria menendezi Gallego & Tassi, 2015	Potrerillos Fm., Middle Triassic, Mendoza, Argentina	Tassi et al., 2015. [41]
Euestheria kozuri Geyer, 2018	Hassberge Fm., Late Triassic, Franconia, Germany	Geyer & Kelber, 2018. [21]
Pseudestherites musacchioi Gallego, 2004 (=*Euestheria* sp.)	La Amarga Fm., Early Cretaceous, Neuquén, Argentina	Musacchio, 1970; Gallego & Shen, 2004. [180,181]
Euestheria sp. undet	New South Wales, Australia	Tasch, 1987. [19]
Euestheria sp.	Longtan Fm., Late Permian, Zhejiang, China	Tang et al., 1980. [182]
Euestheria sp.	Cumnock Fm., Late Triassic, Goldston Quadrangle, Chatham County, USA	Kozur & Weems, 2010. [4]
Euestheria sp.	Sangonghe Fm., Early Jurassic, Xinjiang, China	Fu, 1998. [142]
Euestheria spp.	Middle Permian, Kansas, USA	Tasch, 1958. [105]

Table 1. *Cont.*

Species	Horizon, Age, and Locality	Reference
Euestheria sp.	Albert Fm., New Brunswick, Canada	Greiner, 1974. [110]
Euestheria sp.	Anning Fm., Late Jurassic, Yunnan, China	Cheng et al., 2004. [152]
Euestheria sp.	Canadon Asfalto Fm., Jurassic, Patagonia, South America	Tasch, 1970. [124]
Euestheria spp.	Ershierzhan Fm., Middle Jurassic, Heilongjiang, China	Wang, 1985. [183]
Euestheria ? sp.	Hassberge Fm., Late Triassic, Eltmann, Franconia, Germany	Geyer & Kelber, 2018. [21]
Euestheria spp.	Huangmaqing Fm., Late Triassic, Jiangsu, China	Wu, 1980. [184]
Euestheria sp.	Jurassic, Argentina	Vallati, 1986. [185]
Euestheria spp.	Kayitou Fm., Early Triassic, Guizhou, China	Chu et al., 2013. [78]
Euestheria sp.	Laohugou Fm., Late Triassic, Liaoning, China	Zhang & Dong, 1982. [186]
Euestheria sp.	Late Triassic, Germany	Barth & Kozur et al., 2011. [187]
Euestheria sp.	Tuckahoe Fm., Late Triassic, North Carolina, USA	Kozur & Weems, 2010. [4]
Euestheria sp.	Longfengshan Fm., Middle Jurassic, Gansu, China	Zhang et al., 2013. [149]
Euestheria sp.	Longfenshan Fm., Middle Jurassic, Gansu, China	Li et al., 1982. [141]
Euestheria sp.	Longtan Fm., Late Permian, Zhejiang, China	Tang et al., 1980. [182]
Euestheria sp.	Manantial Pelado Fm., Middle Jurassic, Chubut, Argentina	Gallego, 2010. [188]
Euestheria sp.	Moenave Fm., Early Jurassic, Utah, USA	Milner, 2006. [189]
Euestheria sp.	Pastos Bons Fm., Late Jurassic, Piaui, Brazil	Tasch, 1987. [19]
Euestheria spp.	Rio do Rasto Fm., Late Permian, Santa Catarina, Brazil	Tasch, 1987. [19]
Euestheria spp.	Sangonghe Fm., Early Jurassic, Xinjiang, China	Lu, 1995. [113]
Euestheria sp.	Solling Fm., Triassic, Bavaria, Germany	Kozur & Mock, 1993. [44]
Euestheria sp.	Tiaojishan Fm., Middle Jurassic, Liaoning, China	Gao et al., 2017. [156]
Euestheria sp.	Early Jurassic, Storm Peak, Antarctica	Tasch, 1987. [19]
Euestheria sp.	Wenbinshan Fm., Late Triassic, Fujian, China	Liang, 1998. [135]
Euestheria sp.	Xiaodonggou Fm., Middle Jurassic, Liaoning, Jilin, China	Fan, 1996. [151]
Euestheria sp.	Xingfuzhilu Fm., Early Triassic, Inner Mongolia, China	Zheng et al., 2013. [190]
Euestheria sp.	Xintiangou Fm., Middle Jurassic, Guizhou, China	Zhang et al., 2016. [150]
Euestheria ? sp.	Xishanyao Fm., Middle Jurassic, Xinjiang, China	Lu, 1995. [113]
Euestheria spp.	Xujiashan Fm., Middle Triassic, Jiangsu, China	Min et al., 1981. [138]
Euestheria spp.	Santa Maria Fm., Middle Triassic, Brazil	Katoo, 1971. [26]
Euestheria sp.	Peine Fm., Late Permian, Antofagasta, Chile	Gallego & Breitkreuz, 1994. [174]
Euestheria sp.	Roca Blanca Fm., Early–Middle Jurassic, Santa Cruz, Argentina	Gallego, 1994. [173]
Euestheria sp.	Ischichuca Fm, Middle–Late Triassic, La Rioja, Argentina	Gallego et al., 2001. [51]
Euestheria sp.	Late Triassic, Western Siberia, Russia	Chunikhin, 2004. [178]
Euestheria sp.	Late Cretaceous, Africa	Mouta & Marlière, 1950. [90]
Euestheria sp.	Badaowan Fm., Early Jurassic, Xinjiang, China	Shen, 2003. [140]
Euestheria sp.	Houjiatun Fm., Late Jurassic, Liaoning, China	Shen, 2003. [140]
Euestheria ? sp.	Shanglufeng Fm., Middle Jurassic, Yunnan, China	Zhang et al., 1976. [2]

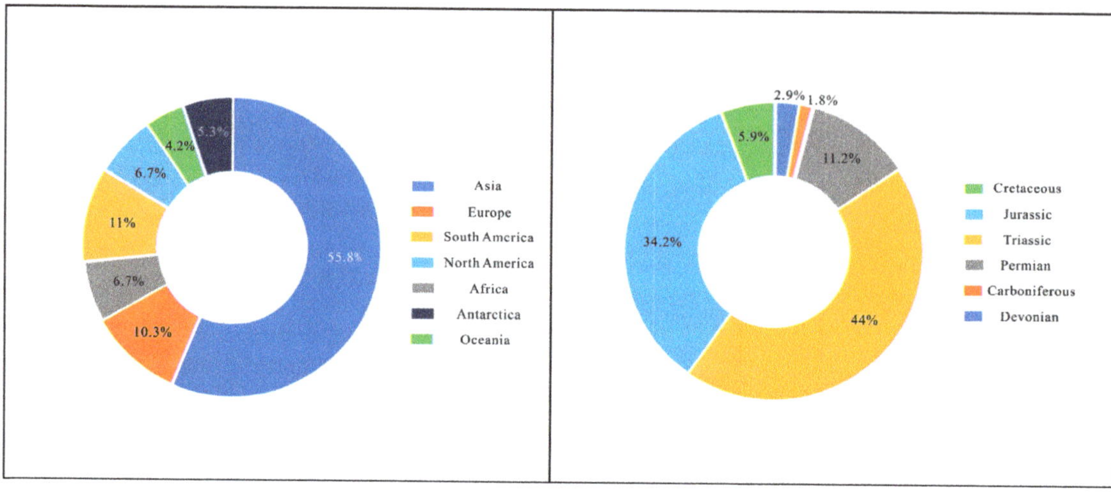

Figure 3. Percentage distribution of *Euestheria* species across modern continents (**left**) and across geological periods according to stratigraphic records (**right**).

2.2. Taxonomic Issues

Most fossil clam shrimps comprise carapaces only (Figure 1c,d). The important features of their crustacean body and appendages are often missing or poorly preserved in fossil specimens. As a result, the key characteristics of the body and appendages that are crucial in the morphological taxonomy of modern clam shrimps could not be used for morphological comparison between the fossil and modern species. The taxonomy of fossil clam shrimps is highly based on its carapace characteristics like the carapace size, shape, growth line numbers, growth band width, larval valve size, and particularly, the micro-ornamentations on the growth bands. The genus *Euestheria* is one of the most common taxa that has been recorded in all seven continents (Figure 3). Its type species *E. minuta* Zieten, 1833, from the Triassic of Germany [191], possesses an oval carapace ornamented with small polygonal reticulations (or irregular lirae) [9,21,117,191]. Observations of fossil clam shrimp in most early studies were performed under optical microscopy, and the micro-ornamentations, especially on those small, badly-preserved specimens, were not generally well recognized. As a result, numerous species with non-reticulate or unknown ornamentations were assigned to *Euestheria* according to their small oval carapace, fine growth lines, and small larval valve, making the genus a "dumping ground" of taxonomy. In recent years, with the widespread use of SEM, an increasing number of euestherids have been found to possess diverse non-reticulate ornamentations that are different from the reticulate ornamentations of the type species. For example, some Jurassic euestherids from China have been amended and reassigned to the genera *Qaidamestheria* Wang, 1983; *Tianzhuestheria* Shen, Li & Chen, 2002; and *Triglypta* Wang, 1984 [94,192–197], and their taxonomic status and descriptions have been changed. These genera possess punctae, punctae-formed reticulations, and linear arrangements or radial lirae on their growth bands [195,196]. Additionally, our recent studies have identified numerous Triassic "euestherids" that do not actually belong to the genus *Euestheria* (separate articles on this topic will be published). *Euestheria* is one of the most common taxa of fossil clam shrimps. However, serious taxonomic issues have impeded the use of this important taxon in global stratigraphic correlation. Hence, further studies on this taxon are desperately needed and will provide great help for comprehending both global stratigraphic correlation and the evolutionary history of clam shrimps.

Author Contributions: Conceptualization, H.L.; methodology, S.F. and H.L.; validation, S.F. and H.L.; formal analysis, H.L.; investigation, S.F. and H.L.; resources, H.L.; data curation, S.F.; writing—original draft preparation, S.F.; writing—review and editing, H.L.; supervision, H.L.; project administration, H.L.; funding acquisition, H.L. All authors have read and agreed to the published version of the manuscript.

Funding: This study was jointly funded by the National Natural Science Foundation of China (42102010), the Open-end Foundation of State Key Laboratory of Palaeobiology and Stratigraphy (183118), and the Yunnan Province Science and Technology Department (202101AT070642).

Data Availability Statement: No new data were created or analyzed in this study. Data sharing is not applicable to this article.

Acknowledgments: We are grateful to Thomas A. Hegna (SUNY Fredonia) for providing a large number of references, and Yaqiong Wang (NIGPAS) for providing specimens of modern spinicaudatans. We are also grateful to two anonymous reviewers for their constructive suggestions.

Conflicts of Interest: The authors declare no conflicts of interest.

References

1. Novojilov, N.I. Novye Phyllopoda iz permskich i triasovych otlozhenij Nordvik–Chatangskogo rajona. *Nedra Arktiki* **1946**, *1*, 172–202. (In Russian)
2. Zhang, W.T.; Chen, P.J.; Shen, Y.B. *Fossil Conchostracans of China*; Science Press: Beijing, China, 1976; 325p. (In Chinese)
3. Chen, P.J.; Shen, Y.B. *An Introduction to Fossil Conchostracans*; Science Press: Beijing, China, 1985; 241p. (In Chinese)
4. Kozur, H.W.; Weems, R.E. The biostratigraphic importance of conchostracans in the continental Triassic of the northern hemisphere. *Geol. Soc. Lond. Spec. Publ.* **2010**, *334*, 315–417. [CrossRef]
5. Ghosh, S.C. Estheriids (Fossil Conchostraca) of Indian Gondwana, Palaeontologia Indica, New Series Number 154. *Geol. Surv. India* **2011**, *54*, 1–290.
6. Gallego, O.F.; Monferran, M.D.; Stigall, A.L.; Zacarías, I.A.; Hegna, T.A.; Jiménez, V.C.; Jiménez, J.S.; Li, G.; Barrios Calathakia, H.G. The Devonian–Cretaceous fossil record of "conchostracans" of Africa and their paleobiogeographic relationships with other Gondwanan faunas. *J. Afr. Earth Sci.* **2020**, *161*, 103648. [CrossRef]
7. Li, G. New spinicaudatan species of Late Jurassic Linglongta Phase of Yanliao Biota from western Liaoning, China. *Zool. Stud.* **2020**, *59*, 36.
8. Liao, H.Y.; Shen, Y.B. An Early Devonian clam shrimp community from Hunan Province, China. *J. Paleontol.* **2022**, *96*, 164–173. [CrossRef]
9. Gross, W. Eine *Estheria* aus dem rheinischen Unterdevon. *Senckenbergiana* **1934**, *16*, 309–313.
10. Maillieux, E. Un conchostracé nouveau de l'assise des grès et schistes de Wépion (Emsien inférieur). *Bull. Musée R. D'histoire Nat. Belg.* **1939**, *15*, 1–7. (In French)
11. Raymond, P.E. The genera of fossil Conchostraca—An order of bivalved Crustacea. *Bull. Mus. Comp. Zool.* **1946**, *96*, 217–307.
12. Defretin, L.S. Crustacés fossiles du Dévonien de l'Artois et du Boulonnais. *Ann. Société Géologique Nord.* **1950**, *70*, 55–66. (In French)
13. Novojilov, N.I. Dvustvorcatye Listonogie Devona. *Tr. Paleontol. Instituta Akad. Nauk* **1961**, *81*, 1–132. (In Russian)
14. Cuvelier, J.; Hennion, M.; Gallego, O.; Vachard, D. Palaeontological collections of Lille University–Sciences and Technologies. II Diplostraca (types and figured specimens). *Ann. Société Géologique Nord.* **2015**, *22*, 119–134. [CrossRef]
15. Hegna, T.A.; Astrop, T.I. The fossil record of the clam shrimp (Crustacea: Branchiopoda). *Zool. Stud.* **2020**, *59*, 43.
16. Kobayashi, T. Fossil estherians and allied fossils. *J. Fac. Sci. Univ. Tokyo Sect. 2 Geol. Mineral. Geogr. Geophys.* **1954**, *9*, 1–192.
17. Molin, V.A.; Novojilov, N.I. *Permian and Triassic Conchostracans from North Russia*; Science Press: Leningrad, Russia, 1965; 117p. (In Russian)
18. Chen, P.J. Handbook of Stratigraphy and Paleontology in Southwest China. In *A Handbook of the Stratigraphy and Palaeontology of Southwest China*; Nanjing Institute of Geology and Paleontology, Chinese Academy of Sciences, Ed.; Science Press: Beijing, China, 1974; pp. 318–320. (In Chinese)
19. Tasch, P. Fossil Conchostraca of the southern hemisphere and continental drift. Paleontology, biostratigraphy and dispersal. *Geol. Soc. Am.* **1987**, *165*, 1–290.
20. Scholze, F.; Golubev, V.K.; Niedźwiedzki, G.; Sennikov, A.G.; Schneider, J.W.; Silantiev, V.V. Early Triassic Conchostracans (Crustacea: Branchiopoda) from the terrestrial Permian–Triassic boundary sections in the Moscow syncline. *Palaeogeogr. Palaeoclimatol. Palaeoecol.* **2015**, *429*, 22–40. [CrossRef]
21. Geyer, G.; Kelber, K.P. Spinicaudata ("Conchostraca", Crustacea) from the Middle Keuper (Upper Triassic) of the southern Germanic Basin, with a review of Carnian–Norian taxa and suggested biozones. *Palz* **2018**, *92*, 1–34. [CrossRef]
22. Scotese, C.R. An Atlas of Phanerozoic Paleogeographic Maps: The Seas Come In and the Seas Go Out. *Annu. Rev. Earth Planet. Sci.* **2021**, *49*, 679–728. [CrossRef]

23. Zieten, C.H.V. Die Versteinerungen Wurttembergs, oder naturgetreue Abbildungen der in den vollsta¨ndigsten Sammlungen, namentlich der in dem Kabinet des Oberamts–Arzt Dr. Hartmann befindlichen Petrefacten, mit Angabe der Gebirgs Formationen, in welchen dieselben vorkommen und der Fundorte. *Unsere Zeit Stuttg.* **1833**, *12*, 89–102.
24. Defretin, L.S. Quelques conchostracés du Nord–Cameroun. *Bull. Serv. Territ. Cameroun* **1953**, *1*, 111–120. (In French)
25. Defretin, L.S. Les conchostracés triassiques du Groenland Oriental. Notes on Triassic stratigraphy and palaeontology of north-eastern Jameson Land (East Greenland). *Meddelelser Om Grønland* **1969**, *168*, 124–136. (In French)
26. Katoo, Y. Conchostráceos Mesozóicos do sul do Brasil: Contribuição à Estratigrafia das Formações Santa Maria e Botucatu. Master's Dissertation, Universidade Federal do Rio Grande do Sul, Porto Alegre, Brazil, 1971; 87p. (In Portuguese).
27. Luo, L.; Wang, D.B.; Chu, D.L.; Wu, Q.J.; Li, K.C. Biostratigraphy and geometric morphometrics of conchostracans from the Middle Triassic in southern lancangjiang zone. *Earth Sci.* **2018**, *43*, 2833–2847. (In Chinese with English Abstract)
28. Gu, Z.W.; Cheng, Z.X. The age division of Triassic Jurassic strata in southwest Hunan. *J. Stratigr.* **1981**, *5*, 241–262+321–322. (In Chinese)
29. Cai, K.D. Continental Triassic system and its fauna from Gansu Province. *Gansu Geol.* **1990**, *10*, 100–120. (In Chinese with English Abstract)
30. Cai, K.D. Triassic system in Gansu. *Acta Gansu Geol.* **1993**, *1*, 50–100+135–136. (In Chinese with English Abstract)
31. Gallego, O.F. Conchóstracos triásicos de Mendoza y San Juan, Argentina. *Ameghiniana* **1992**, *29*, 159–175. (In Spanish)
32. Gallego, O.F. *Estudio Sistemático de las Faunas de Conchóstracos Triásicos de la República Argentina*; Universidad Nacional de Córdoba: Córdoba, Argentina, 1999; pp. 1–210. (In Spanish)
33. Gallego, O.F. *Tryasoglypta santamariensis* Gallego nov. comb. (Conchostraca) de la Formación Santa María (Triásico Medio–Superior) de Brasil. *Geociências* **1999**, *4*, 61–66. (In Spanish)
34. Knox, L.W.; Gordon, E.A. Ostracodes as indicators of brackish water environments in the Catskill Magnafacies (Devonian) of New York State. *Palaeogeogr. Palaeoclimatol. Palaeoecol.* **1999**, *148*, 9–22. [CrossRef]
35. Zhang, W.; Lin, H.W. Mesozoic and Paleogene conchostracan assemblages of Guangdong with reference to the origin of Dongjiang and Xijiang Rivers. *J. Stratigr.* **2000**, *24*, 300–302. (In Chinese with English Abstract)
36. Kozur, H.W.; Weems, R.E. Upper Triassic conchostracan biostratigraphy of the continental rift basins of eastern North America: Its importance for correlating Newark Supergroup events with the Germanic Basin and the international geologic time scale. In *The Global Triassic*; Lucas, S.G., Spielmann, J.A., Eds.; New Mexico Museum of Natural History and Science Bulletin: Mexico City, Mexico, 2007; pp. 137–188.
37. Sun, G.; Miao, Y.; Mosbrugger, V.; Ashraf, A.R. The Upper Triassic to Middle Jurassic strata and floras of the Junggar Basin, Xinjiang, Northwest China. *Palaeobiodivers. Palaeoenviron.* **2010**, *90*, 203–214. [CrossRef]
38. Jenisch, A.G.; Lehn, I.; Gallego, O.F.; Monferran, M.D.; Horodyski, R.S.; Faccini, U.F. Stratigraphic distribution, taphonomy and paleoenvironments of Spinicaudata in the Triassic and Jurassic of the Paraná Basin. *J. S. Am. Earth Sci.* **2017**, *80*, 569–588. [CrossRef]
39. Wang, S.E.; Liu, S.W. Fossil conchostracans. In *Mesozoic Strata and Palaeontology of the Shaanxi–Gansu-Ningxia Basin, Part 2*; Institute of Geology, Chinese Academy of Geological Sciences, Ed.; Geological Publishing House: Beijing, China, 1980; pp. 84–110. (In Chinese)
40. Shen, Y.B.; Chen, P.J. Conchostraca. In *Upper Triassic Stratigraphy, Palaeontology and Palaeogeography Along the Northern Circum Pacific Belt of China*; Mi, J., Zhang, C., Sun, C., Eds.; Science Press: Beijing, China, 1993; pp. 181–183. (In Chinese)
41. Tassi, L.V.; Zavattieri, A.M.; Gallego, O.F. Triassic spinicaudatan fauna from the Cerro de Las Cabras Formation (Cuyo Basin), Mendoza Province (Argentina): Description of new species and revision of previous records. *Ameghiniana* **2015**, *52*, 241–264. [CrossRef]
42. Voltz, M. Notice sur le Grès Bigarré de la grande carrière de Soultz-Les-Bains. *Mémoires Société Muséum D'histoire Nat. Strasbg.* **1835**, *2*, 1–14.
43. Lucas, S.G.; Schoch, R.R. Triassic temnospondyl biostratigraphy, biochronology and correlation of the German Buntsandstein and North American Moenkopi Formation. *Lethaia* **2002**, *35*, 97–106. [CrossRef]
44. Kozur, H.W.; Mock, R. The importance of conchostracans for the correlation of continental and marine beds. The Nonmarine Triassic. *New Mex. Mus. Nat. Hist. Sci. Bull.* **1993**, *3*, 261–266.
45. Lea, I. On fossils from the New Red Sandstone Formation of Pennsylvania. *Proc. Acad. Nat. Sci. Phila.* **1856**, *8*, 77–78.
46. Spamer, E.E. Notes on six real and supposed type fossils from the Newark Supergroup (Triassic) of Pennsylvania. *Mosasaur* **1989**, *4*, 49–52.
47. Jones, T.R. A monograph of the fossil Estheriae. *Monogr. Palaeontogr. Soc.* **1862**, *14*, 1–134. [CrossRef]
48. Pang, Q. The nonmarine Triassic and Ostracoda in northern China. *New Mex. Mus. Hist. Sci. Bull.* **1993**, *3*, 383–392.
49. Kozur, H.W.; Weems, R.E. Additions to the uppermost Alaunian through Rhaetian (Triassic) conchostracan zonation of North America. *Foss. Rec.* **2011**, *3*, 295–300.
50. Liu, S.W.; He, Z.H. Marine conchostracans from the "Sunjiagou Formation" of Qishan, Shaanxi. *Acta Palaeontol. Sin.* **2000**, *39*, 230–236. (In Chinese with English Abstract)
51. Gallego, O.F.; Melchor, R.N.; Chalpe, V. Stratigraphic and paleoenvironmental distribution of the conchostracan fauna from the Ischichuca Formation (Triassic), La Rioja, Argentina. *Ameghiniana* **2001**, *38* (Suppl. S4), 33R.

52. Gallego, O.F. Revisión de algunos conchóstracos de la Formación Santa María (Triásico Medio a Superior) del Estado de Rio Grande del Sur (Brasil). *Acta Geol. Leopoldensia* **1996**, *19*, 59–76. (In Spanish)
53. Eichwald, E. *Lethaea rossica ou Paléontologie de la Russie: 2: Second Volume. Premiere Section de la Periode Moyenne, Contenant les Vegetaux, les Spongiaires, les Coraux, les Radiaires, les Annelides, et le Commencement des Mollusques*; E. Schweizerbart: Stuttgart, Germany, 1868; 678p. (In French)
54. Jones, T.R. On some bivalved Entomostraca from the coal-measures of South Wales. *Geol. Mag.* **1870**, *7*, 214–220. [CrossRef]
55. Copeland, M.J. The arthropod fauna of the Upper Carboniferous rocks of the Maritime Provinces. *Geol. Surv. Can. Mem.* **1957**, *286*, 1–110.
56. Bell, W.A. Mississippian Horton Group of type Windsor–Horton District, Nova Scotia. *Geol. Surv. Can. Mem.* **1960**, *314*, 1–112.
57. Goldenberg, F. Die fossilen Thiere aus der Steinkohlenformation von Saarbrücken. *Fauna Saræpontana Foss.* **1877**, *2*, 1–54. (In German)
58. Jones, T.R. Notes on some bivalved Entomostraca. *Geol. Mag.* **1878**, *3*, 100–102. [CrossRef]
59. Jones, T.R.; Etheridge, R.; Woodward, H. Second report of the committee, consisting of Mr. R. Etheridge, Dr. H. Woodward, and Professor T. Rupert Jones (Secretary), on the fossil Phyllopoda of the Palaeozoic rocks. *Rep. Br. Assoc. Adv. Sci.* **1883**, *54*, 75–95.
60. Jones, T.R. On some fossil Estheriae. *Geol. Mag. New Ser.* **1890**, *7*, 385–390. [CrossRef]
61. Olempska, E. Late Triassic spinicaudatan crustaceans from southwestern Poland. *Acta Palaeontol. Pol.* **2004**, *49*, 429–442.
62. Grand, E. Flore carbonifère de la Loire et du Centre de la France. *Mémoire L'académie Des Sci.* **1890**, *24*, 124–154. (In French)
63. Jones, T.R. On the Enon conglomerate of the Cape of Good Hope and its fossil Estheriae. *Geol. Mag.* **1901**, *8*, 350–354. [CrossRef]
64. Pruvost, P. Notes sur les entomostracés bivalves du terrain houiller du Nord de la France. *Ann. Soc. Geol. Nord* **1911**, *40*, 60–80. (In French)
65. Dix, E.; Trueman, A.E. Marine Horizons in the Coal Measures of South Wales. *Geol. Mag.* **1928**, *65*, 356–363. [CrossRef]
66. Schmidt, M. *Die Lebewelt Unserer Trias. Nachtrag 1938*, Hohenlohe'sche Buchghandlung Ferdinand Rau: Oehringen, Germany, 1938; pp. 1–144.
67. Ozawa, W. On Two New Species of *Estheria* from the Mesozoic shale of Korea. *Jpn. J. Geol. Geogr.* **1923**, *2*, 40–42.
68. Kim, J.H.; Lee, G.H. Fossil Conchostraca from the Amisan Formation of the Nampo Group, Korea. *J. Korean Earth Sci. Soc.* **2015**, *36*, 181–189. (In Korean with English Abstract) [CrossRef]
69. Tchernychev, B. Sur Estheria et Estheriella du bassin de Donetz. *Ezhegodnik Vsesoiuznogo Paleontol. Obs.* **1926**, *6*, 67–82. (In Russian)
70. Mitchell, J. The fossil Estheriae of Australia, part 1. *Proc. Linn. Soc. New South Wales* **1927**, *52*, 105–112.
71. Chernyshev, B.I. Estheria from Siberia and the Far East. *Cent. Geol. Prospect. Bur. Bull.* **1930**, *49*, 85-70.
72. Ji, R.S. On the occurrence of fossil estherian China, and its geology significance. *J. Geol. Soc. China* **1931**, *1*, 189–223+225–227+361.
73. Wei, J.M. Conchostracan. In *Fossil Atlas of Northwest China, Xinjiang Uygur Autonomous Region: III Mesozoic and Cenozoic*; Geological Survey of Xinjiang Administrative Bureau of Petroleum, Regional Surveying Team of Xingjiang Geological Bureau, Ed.; Geological Publishing House: Beijing, China, 1984; pp. 108–112. (In Chinese)
74. Poschmann, M.J.; Hegna, T.A.; Astrop, T.I.; Hoffmann, R. Revision of Lower Devonian clam shrimp (Branchiopoda, Diplostraca) from the Rhenish Massif (Eifel, SW-Germany), and the early colonization of non-marine palaeoenvironments. *Palaeobiodivers. Palaeoenviron.* **2024**, *104*, 535–569. [CrossRef]
75. Lyutkevich, E.M. Onekotorych Phyllopoda SSSR. *Ezhegodnik Vsesoyuznogo Paleontol. Obs.* **1937**, *11*, 60–70. (In Russian)
76. Kozur, H.W.; Seidel, G. Revision der Conchostracen–Faunen des unteren und mittleren Buntsandsteins Teil I. *Earth Sci.* **1983**, *11*, 289–417.
77. Ptaszyński, T.; Niedźwiedzki, G. Conchostraca (muszloraczki) z najniższego pstrego piaskowca Zachełmia, Góry Świętokrzyskie. *Przegląd Geol.* **2004**, *52*, 1151–1155. (In Polish)
78. Chu, D.L.; Tong, J.N.; Yu, J.X.; Song, H.J.; Tian, L. The conchostracan fauna from the Kayitou Formation of western Guizhou, China. *Acta Palaeontol. Sin.* **2013**, *52*, 265–280. (In Chinese with English Abstract)
79. Chu, D.L.; Miao, X.; Wu, Y.S.; Guo, W.W.; Shu, W.C.; Tong, J.N. Conchostracans from the Permian–Triassic Transition in Weibei Area of Shaanxi Province and Its Biostratigraphic. *Earth Sci.* **2018**, *43*, 3910–3921. (In Chinese with English Abstract)
80. Scholze, F.; Shen, S.Z.; Backer, M.; Wei, H.B.; Hübner, M.; Cui, Y.Y.; Feng, Z.; Schneider, J.W. Reinvestigation of conchostracans (Crustacea: Branchiopoda) from the Permian–Triassic transition in southwest China. *Palaeoworld* **2020**, *29*, 368–390. [CrossRef]
81. Miao, X.; Chu, D.L.; Tong, J.N.; Benton, M.J.; Yu, J.X.; Luo, L.; Shu, W.C.; Wu, Y.Y. Biostratigraphic significance and geometric morphometrics of *Euestheria gutta* (Crustacea: Branchiopoda: Spinicaudata): An index fossil of continental Permian–Triassic transitional beds. *Geol. J.* **2021**, *56*, 6176–6188. [CrossRef]
82. Scholze, F.; Hamad, A.A.; Schneider, J.W.; Golubev, V.K.; Sennikov, A.G.; Dieter Uhl, S.V. An enigmatic 'conchostracan' fauna in the eastern Dead Sea region of Jordan: First records of Rossolimnadiopsis Novojilov from the Early Triassic Main Formation. *Palaeogeogr. Palaeoclimatol. Palaeoecol.* **2017**, *466*, 314–325. [CrossRef]
83. Vladimirovna, Z.V.; Frank, S.; Ivanovich, D.V. Early Triassic conchostracans from the Tiryakh-Kobyume section (Southern Verkhoyansk region, Republic of Sakha (Yakutia)). *Uchenye Zapiski Kazanskogo Universiteta. Seriya Estestvennye Nauki.* **2020**, *162*, 244–252.
84. Schmidt, M. *Die Lebewelt Unserer Trias*; Hohenlohe'sche Buchghandlung Ferdinand Rau: Oehringen, Germany, 1928; pp. 1–461.

85. Defretin, L.S. Sur quelques *Estheria* du Trias français à facies germanique et de l'Hettangien. *Ann. Société Géologique Nord* **1950**, *70*, 214–227. (In French)
86. Novojilov, N.I. Conchostraca de la super famille des Limnadiopseidea superfam. nov. *Ann. Serv. D'information Géologique Bur. Rech. Géologique Géophysiques Miniéres* **1958**, *26*, 95–128. (In Russian)
87. Deleau, P. Les bassins carboniferes du Sud–Oranais (Algerie); le bassin de Colomb–Bechar–Kenadza et le bassin du Guir. *Bull. Société Géologique Fr.* **1945**, *5*, 625–632. (In French) [CrossRef]
88. Teixeira, C. Acerca des filopodes fósseis do Karrov da Escarpa do Quela (Angola): Lisboa. *Estud. Geol. Paleontol. An.* **1947**, *2*, 29–43.
89. Rusconi, C. Algunas especies de esterias del Triasico en Mendoza. *Rev. Del Mus. Hist. Nat. Mendoza* **1948**, *2*, 199–202. (In Spanish)
90. Mouta, F.; Marlière, R. As camadas a filópode da série de Cassanje (Karroo) no norte de Angola. *An. Estud. Geol. Paleontol. Junta Das Missões Geográficas Investig. Colon.* **1950**, *5*, 11–76. (In Russian)
91. Egoroff, A.; Lombard, A.L. Présence des couches de Stanleyville dans le sous–sol de Léopoldville, République de Congo. note préliminaire. *Ann. Société Géologique Belg.* **1961**, *85*, 103–109. (In French)
92. Almeida, F.F.M. Una faunula de Crustáceos bivalvos do Arenito Botucatú no Estado de São Paulo. Ministério da Agricultura. Departamento Nacional da Produção Mineral. *Div. Geol. Mineral. Bol.* **1950**, *134*, 7–38.
93. Kobayashi, T. Older Mesozoic *Estherites* from Eastern Asia. *J. Fac. Sci. Tokyo Univ.* **1951**, *7*, 431–440.
94. Wang, S.E.; Liu, S.W.; Niu, S.W. *Conchostraca. Paleontological Atlas of North China 2, Mesozoic*; Institute of Geology, Chinese Academy of Geological Sciences, Ed.; Science Press: Beijing, China, 1984; pp. 72–123. (In Chinese)
95. Bock, W. American Triassic estherids. *J. Paleontol.* **1953**, *27*, 62–76.
96. Gallego, O.F.; Shen, Y.B.; Cabaleri, N.G.; Hernández, M. The genus *Congestheriella* Kobayashi, 1954 ("Conchostraca", Diplostraca, Afrograptioidea): Redescription and new combination to *Isaura olsoni* Bock from Venezuela and a new species from Argentina (Upper Jurassic). *Alavesia* **2010**, *3*, 11–24.
97. Weems, R.E.; Lucas, S.G. A revision of the Norian conchostracan zonation in North America and its implications for Late Triassic North American tectonic history. *New Mex. Mus. Nat. Hist. Sci. Bull.* **2015**, *67*, 303–317.
98. Alarcón, C.M.; Colombi, C.E.; Gallego, O.F.; Drovandi, J.M.; Drovandi, M.D. Dispersal of Late Triassic clam shrimps across Pangea linking northwestern Gondwana and central Pangea rift basins. *Sci. Rep.* **2024**, *14*, 15025. [CrossRef]
99. Novojilov, N.I. Bivalve phyllopod crustaceans of the Devonian of the Kursk region. *Dokl. Akad. Nauk SSSR* **1953**, *91*, 947–949. (In Russian)
100. Stepanov, I.V. About new finds of conchostracans from Mesozoic continental deposits of the Irkutsk and Kansk–Achinsk coal basins. In *Biostratigraphy and Paleontology of the Continental Upper Paleozoic and Mesozoic of Siberia and Central Kazakhstan (Biostratigraphy and Paleontology)*; Akademiâ Nauk, S.S.S.R., Ed.; Science Press: Leningrad, Russia, 1966; pp. 150–183. (In Russian)
101. Mendes, J.C. Conchostracos permianos do Sul do Brasil. In *Paleontologia do Paraná*; Lange, F.W., Ed.; Volume Comemorativo do 1.°; Centenário do Estado do Paraná: Paraná, Brazil, 1954; pp. 153–164.
102. Novojilov, N.I. Characteristic complex of the fauna of the Ilemorovo Formation (D22 ilem). Class Crustacea. In *Field Atlas of Characteristic Fauna and Flora Complexes of the Devonian Deposits of the Minusinsk Basin*; Rzhonsnitskaya, M.A., Meleshchenko, V.A., Eds.; Gosudarstvennoye Nauchno–Tekhnicheskoye Izdatel'stvo: Moscow, Russia, 1955; pp. 31–33. (In Russian)
103. Novojilov, N.I. Characteristic fauna complex of the Kohai Formation (D31 koch). Class Crustacea. In *Field Atlas of Characteristic Fauna and Flora Complexes of the Devonian Deposits of the Minusinsk Basin*; Rzhonsnitskaya, M.A., Meleshchenko, V.A., Eds.; Gosudarstvennoye Nauchno–Tekhnicheskoye Izdatel'stvo: Moscow, Russia, 1955; pp. 43–45. (In Russian)
104. Tasch, P. Three general principles for a system of classification of fossil conchostracans. *J. Paleontol.* **1956**, *30*, 1248–1257.
105. Tasch, P. Permian conchostracan–bearing beds of Kansas, Pt. 1. Jester Creek Section: Fauna and paleoecology. *J. Paleontol.* **1958**, *32*, 525–540.
106. Defretin, L.S.; Joulia, F.; Lapparent, A.F. Les *Estheria* de la région d'Agades (Niger). *Bull. Société Géologique Fr.* **1956**, *6*, 679–690. (In French) [CrossRef]
107. Defretin, L.S. Remarques apropos de la note de NI Novojilov sur quelques Conchostracés chinois et africain. *Ann. Société Géologique Nord* **1958**, *77*, 244–260. (In French)
108. Pinto, I.D. Artropodos da Formaçao Santa Maria (Triassico Superior) do Rio Grande do Sul, com noticias sobre alguns restos vegetais. *Bol. Soc. Bras. Geol.* **1956**, *5*, 76–87.
109. Kashirtsev, A.S. On the Upper Paleozoic fauna of western Verkhoyania and its stratigraphic significance. *Vopr. Paleobiogeografii I Biostratigrafii* **1957**, *12*, 179–190. (In Russian)
110. Greiner, H. The Albert Formation of New Brunswick: A Paleozoic lacustrine model. *Geol. Rundsch.* **1974**, *63*, 1102–1113. [CrossRef]
111. Kusumi, H. On the occurrence of Cretaceous Estheriids in North Kyushu. Journal of Science of the Hiroshima University. *Ser. C Geol. Mineral.* **1960**, *3*, 15–24.
112. Zaspelova, V.S. Phyllopoda and Ostracoda from lower Mesozoic deposits of Kongel syncline. *Trans. Coal Geol. Lab. Acad. Sci. USSR* **1961**, *12*, 218–231. (In Russian)
113. Lu, H.N. The Jurassic of the Junggar Basin, Xinjiang. *J. Stratigr.* **1995**, 180–190. (In Chinese with English Abstract)
114. Chen, P.J. Jurassic conchostracans from Mengyin District, Shandong. *Acta Palaeontol. Sin.* **1982**, *21*, 135–139. (In Chinese with English Abstract)

115. Chen, J.H. Jurassic strata and paleogeographic outline. In *Stratigraphy, Paleogeography and Plate Tectonics of NW China*; Zhou, Z.Y., Lin, H.L., Eds.; Nanjing University Press: Nanjing, China, 1995; pp. 253–265.
116. Chen, P.J. Jurassic biostratigraphy of China. In *Biostratigraphy of China*; Zhang, W.T., Chen, P.J., Palmer, A., Eds.; Science Press: Beijing, China, 2003; pp. 423–463.
117. Reible, P. Die Conchostraken (Branchiopoda, Crustacea) der Germanischen Trias. Neues Jahrbuch für Geologie und Paläontologie. *Abhand-Lungen* **1962**, *114*, 169–244.
118. Sell, J. The Euestheriidae (Conchostraca) of the Upper Muschelkalk and Lower Keupers of Lower Franconia. *Semana* **2018**, *33*, 55–90.
119. Molin, V.A. Bivalve phyllopods from the Lower Triassic of eastern Yakutia. In *Paleontology and Biostratigraphy of the Paleozoic and Triassic Deposits in Yakutia*; Vozin, V.F., Ed.; Science Press: Leningrad, Russia, 1965; pp. 91–114. (In Russian)
120. Wu, T.Y. Conchostracan assemblage from bottom of Ermayin Formation, Shaanxi. *Acta Palaeontol. Sin.* **1991**, *30*, 630–636+641–642+678–680. (In Chinese with English Abstract)
121. Molin, V.A. New Lower Triassic phyllopods of Pechora and Mezen. In *Stratigrafiâ i Paleontologiâ Severo–Vostoka Evropejskoj Časti SSSP*; Akademiâ Nauk, S.S.S.R., Ed.; Science Press: Leningrad, Russia, 1966; pp. 62–73. (In Russian)
122. Lipatova, V.V.; Lopato, A.Û. *Triassic Phyllopoda (Crustacea) of Eurasia and Their Stratigraphic Value*; Geos: Moscow, Russia, 2000; 124p. (In Russian)
123. Defretin, L.S. Étude sur les phyllopodes du Bassin du Congo. *Ann. Mus. R. L'afrique Cent. Sci. Geol.* **1967**, *56*, 1–122. (In French)
124. Tasch, P.; Volkheimer, W. Jurassic conchostracans from Patagonia. *Univ. Kans. Paleontol. Contrib.* **1970**, *50*, 1–23.
125. Novojilov, N.I. *Vymershie limnadioidei [Extinct Limnadioidea]*; Nauka: Moscow, Russia, 1970; 249p. (In Russian)
126. Chunikhin, S.A. Conchostracans of the Permain and Triassic of western Siberia. *Tyumen State Oil Gas Univ. Tomsk.* **2009**, *25*, 1–149. (In Russian)
127. Zaspelova, V.S.; Žigajte, V.K.; Molin, V.A.; Stepanov, I.V. New Paleozoic and Early Mesozoic conchostracans of USSR and Svalbard. In *New Species of Ancient Plants and Invertebrates of USSR*; Zanina, I.E., Ed.; Izdatel'stvo Nauka: Moskva, Russia, 1972; pp. 247–254. (In Russian)
128. Chen, P.J.; Shen, Y.B. Upper Triassic conchostracans from Qamdo Region. In *Palaeonatology of Xizang III*; Chinese Academy of Sciences Tibetan Plateau Comprehensive Scientific Expedition Team, Ed.; Science Press: Beijing, China, 1981; pp. 341–344. (In Chinese with English Abstract)
129. Kobayashi, T. Upper Triassic estheriids in Thailand and the conchostracan development in Asia in the Mesozoic Era. *Contrib. Geol. Palaeontol. Southeast Asia CLXV Geol. Palaeontol. Southeast Asia* **1975**, *16*, 57–90.
130. Chitnarin, A.; Kershaw, S.; Promduang, A.; Tepnarong, P. Late Triassic freshwater conchostracan, ostracods, and stromatolites from Huai Hin Lat Formation, northeastern Thailand. *Thai Geosci. J.* **2022**, *3*, 32–50.
131. Chonglakmani, C.; Duan, W.W.; Fontaine, H. Note on the continental deposits of peninsular Thailand with a description of some conchostracans. *Oil Gas Geol.* **1990**, *2*, 31–42.
132. Kobayashi, T. Mesozoic Conchostraca of Thailand and Indonesia with notes on conchostracan palaeontology. *Geol. Palaeontol. Southeast Asia* **1984**, *25*, 285–288.
133. Burrett, C. The Man Who Made Impact–The Scientific Work of Dr Sangad Bunopas. *Thai Geosci. J.* **2021**, *2*, 1–15.
134. Lv, B.Y.; Hu, S.B.; Zhu, X.Z.; Pu, T. Stratigraphic characteristics of the Maichuqing Formation of the Triassic Maichuqing Formation in Lagu Houshan, Nanjian County. *Technol. Mark.* **2016**, *23*, 49–50. (In Chinese)
135. Liang, S.J. Sequence and Characters of conchostracan fauna in Fujian Province. *Geol. Fujian* **1998**, *2*, 57–67. (In Chinese with English Abstract)
136. Cao, B.S.; Liang, S.J.; Zhang, Z.M.; Zhang, X.Q.; Ma, A.S. A preliminary study on Jurassic biostratigraphy in Fujian Province. *Geol. Fujian* **1989**, *3*, 198–216+250–255+258–259. (In Chinese with English Abstract)
137. Kozur, H.W. Die Conchostraken–Fauna der mittleren Bernburg–Formation (Buntsandstein) und ihre stratigraphische Bedeutung. *Z. Für Geol. Wiss.* **1980**, *7*, 885–903.
138. Min, Q.K.; Zhou, G.X.; Jiang, L.F.; Chen, C.Z.; Wen, S.X.; Zhang, Z.M. A restudy of the Triassic system in Nanjing area. *J. Stratigr.* **1981**, *1*, 1–9. (In Chinese)
139. Li, C.Y. A discussion on the age of Dongyingfang Formation in Tianshifu District from fossil Conchostraca. *Liaoning Geol.* **1990**, *1*, 42–49. (In Chinese with English Abstract)
140. Shen, Y.B. Conchostracan fauna. In *Jurassic in the North of China, 1*; Deng, S.H., Chen, P.J., Eds.; Petroleum Industry Press: Beijing, China, 2003; pp. 50–58. (In Chinese)
141. Li, B.X.; Xu, F.X.; Mar, C.H.; Pan, H.Z.; Wang, S.Q.; Li, Z.W. Middle Jurassic stratigraphy in Wangjiashan Basin, Jingyuan, Gansu. *J. Stratigr.* **1982**, *1*, 33–40. (In Chinese)
142. Fu, J.H. Early Jurassic conchostracans on the southern margin of Junggar Basin, Xinjiang. *J. Stratigr.* **1998**, *22*, 129–131.
143. Feng, R.L. Crustacea. In *Fossil Atlas of Southwestern China II: Carboniferous to Quaternary*; Work Team of Stratigraphy and Palaeontology of Guizhou Province, Ed.; Geological Publishing House: Beijing, China, 1978; pp. 445–452.
144. Huang, Q.S.; Lu, Z.H.; Huang, J.Y.; Peng, W.S.; Zhang, D.X. The New Discovery and Classification Program of the Early and Middle Jurassic Strata in the Northeast Jiangxi. *J. Stratigr.* **1998**, *22*, 67–74. (In Chinese with English Abstract)
145. Shen, Y.B.; Chen, P.J.; Huang, D.Y. Age of the fossil conchostracans from Daohugou of Ningcheng, Inner Mongolia. *J. Stratigr.* **2003**, *27*, 311–313+373. (In Chinese with English Abstract)

146. Shen, Y.B. Conchostracan. In *Cretaceous and Jurassic Stratigraphy and Environment of the Chaoshui and Yabrai Basins, NW China*; Nanjing Institute of Geology and Palaeontology, Ed.; University of Science and Technology of China Press: Hefei, China, 2004; pp. 54–55. (In Chinese)
147. Bao, Y.W.; Su, M.R.; Tan, Q.; Bao, F.Q. Revision of Middle Jurassic Wanbao Formation around China Mongolia boroer in the west of Manzhouli, Inner Mongolia. *Geol. Resour.* **2011**, *20*, 12–15. (In Chinese with English Abstract)
148. Zhang, Y.M.; Lü, X.Y.; Hou, J.; Ma, W.; Wang, R.; Pan, S.J. Stratum division contrast and geologic age of Tuchengzi Formation–Yixian Formation in western part of Liaoning Province. *Jilin Geol.* **2012**, *31*, 1–6. (In Chinese with English Abstract)
149. Zhang, Y.K.; Li, Z.; Liu, M.Q.; Yu, J.P. Discovery and significance of Middle Jurassic fossils in Jiutiaoling Basin, Middle Qilian Mountains. *Gansu Sci. Technol.* **2013**, *29*, 36–38. (In Chinese)
150. Zhang, X.S.; Zhao, B.; Tan, M.; Zhou, B.Y.; Sun, J. Stratigraphic characteristics of Ziliujing Formation, Jurassic series and discovery of dinosaur footprints in Dafang, Guizhou. *Guizhou Geol.* **2016**, *33*, 50–57+70. (In Chinese with English Abstract)
151. Fan, G.Q. On the problem about Middle Jurassic Ic–Early cretaceous strata from East Liaoning to South Jilin. *Liaoning Geol.* **1996**, *2*, 19–29. (In Chinese with English Abstract)
152. Cheng, Z.W.; Li, P.X.; Pang, Q.Q.; Zhang, Z.X.; Zhang, Z.J.; Jin, Y.G.; Lu, L.W.; Fang, X.G. New progress in the study of the Jurassic of central Yunnan. *Geol. Bull. China* **2004**, *2*, 154–159. (In Chinese with English Abstract)
153. Liao, H.Y. *Biostratigraphy of Conchostracans in the Middle–Late Jurassic Yanliao Biota*; Nanjing Institute of Geology and Paleontology, Chinese Academy of Sciences: Nanjing, China, 2017; pp. 1–169. (In Chinese with English Abstract)
154. Tan, W.; Song, J.; Wang, J.; Yang, H.; Wang, R.; Yu, X. Biological characteristics and time of the Hadataolegai Formation in the central Daxinganling Mountains. *Eng. Constr.* **2019**, *2*, 55. (In Chinese)
155. Peng, L.; Yang, T.; Liu, J.; Chen, Y.Q. Two Middle Jurassic *Euestheria* species fossils in Baojishan Basin, Gansu Provive. *Gansu Geol.* **2016**, *25*, 1–6+49. (In Chinese with English Abstract)
156. Gao, F.L.; Jiang, Y.; Zhang, G.R.; Pan, Y.Q.; Wu, Z.J.; Wang, X. Location and new found of Yanliao Biota in Western Liaoning. *Geol. Rev.* **2017**, *63*, 770–780. (In Chinese with English Abstract)
157. Liu, S.W. Conchostracans in Baicheng District of Xinjiang. *Xinjiang Geol.* **1990**, *8*, 176–184+193–195. (In Chinese with English Abstract)
158. Chen, Q.S.; Ma, W.P.; Cao, Z.Y.; Chen, P.J.; Shen, Y.B.; Lin, Q.B. Time problems of coal-bearing strata in Wuzao Formation, Zhejiang. *Acta Stratigr. Sin.* **1978**, *2*, 76–80. (In Chinese)
159. Duan, W.W. Conchostraca. In *Paleontological Atlas of Southwestern China, Sichuan Province: II*; Southwest Institute of Geological Sciences , Ed.; Geological Publishing House: Beijing, China, 1978; pp. 448–460. (In Chinese with English Abstract)
160. Wu, T.Y. Middle Jurassic conchostracans from Shiguaizi Coal–Field, Inner Mongolia. *Acta Palaeontol. Sin.* **1980**, *19*, 340–342+360. (In Chinese with English Abstract)
161. Hu, B. Late Triassic–Middle Jurassic continental stratigraphy in Jiyuan, Henan Province. *J. Stratigr.* **1991**, *15*, 48–52. (In Chinese with English Abstract)
162. Xu, J.F. Mesozoic conchostracans from some localities in southwest Sichuan Basin. In *Continental Mesozoic Stratigraphy and Paleontology in Sichuan Basin of China*; Part II. Paleontological Professional, Papers; Jurassic and Cretaceous Conchostracans from Sichuan Basin, Ed.; Sichuan People's Publishing House: Chengdu, China, 1982; pp. 416–424. (In Chinese)
163. Shen, Y.B.; Garassino, A.; Teruzzi, G. Studies on Permo–Trias of Madagascar. 4. Early Triassic conchostracans from Madagascar. *Atti Della Soc. Ital. Sci. Nat. Del Mus. Civ. Stor. Nat. Milano* **2002**, *143*, 3–11.
164. Tasch, P. Biostratigraphy and paleontology of some conchostracan–bearing beds in southern Africa. *Paleontol. Afr.* **1984**, *15*, 61–85.
165. Lu, W.Y.; Li, Y.F.; Zhou, G.X.; Cheng, C.D.; Yao, G.H.; Shen, Y.B.; Cao, Z.Y.; Lin, Q.B.; Li, W.B. Jurassic in Chaohu area, Anhui. *J. Stratigr.* **1985**, 180–185. (In Chinese)
166. Zhang, W.T.; Shen, Y.B.; Niu, S.W. Discovery of Jurassic conchostracans with well–preserved soft parts and notes on its biological significance. *Acta Palaeontol. Sin.* **1987**, *26*, 111–129+223–228. (In Chinese with English Abstract)
167. Cao, B.S. Early–Middle Jurassic conchostracans from southwestern Fujian. *Acta Palaeontol. Sin.* **1986**, *25*, 328–332+374. (In Chinese)
168. Carvalho, I.S. *Os Conchostráceos Fósseis das Bacias Interiores do Nordeste do Brasil. Rio de Janeiro*; Tese de Doutorado, Programa de Pós-Graduação em Geologia, Universidade Federal do Rio de Janeiro: Rio de Janeiro, Brazil, 1993; 319p.
169. Cardoso, R.N. Contribuição ao estudo da Formação Areado: Estratigrafia e descrição dos filópodos fósseis. *Arq. Mus. História Nat.* **1971**, *1*, 9–47.
170. Chen, P.J.; Hudson, J.D. The conchostracan fauna of the great estuarine group, Middle Jurassic, Scotland. *Palaeontology* **1991**, *34*, 515–545.
171. Zhang, Y.Z.; Li, G.; Teng, X.; Wang, L.H.; Cheng, X.S. New Jurassic spinicaudatans from Xinjiang Uygur Autonomous Region of northwestern China and their evolutionary implication. *Palaeoworld* **2017**, *26*, 663–671. [CrossRef]
172. Gallego, O.F.; Herbst, R.; Ferrando, L.A. *Cyzicus (E.) falconeri n.* sp. (Conchostraca) de la Formación Yaguarí (Pérmico Superior), Uruguay. *Ameghiniana* **1993**, *30*, 17–22. (In Spanish)
173. Gallego, O.F. Conchóstracos (Cyzicidae) del Jurásico de Santa Cruz y Chubut, Argentina. *Ameghiniana* **1994**, *31*, 333–345. (In Spanish)
174. Gallego, O.F.; Breitkreuz, C. Conchostracos (Crustaceae–Conchostraca) paleozoicos de la Region de Antofagasta, norte de Chile. *Rev. Geol. Chile* **1994**, *21*, 31–53. (In Spanish)
175. Gallego, O.F.; Campos, C.C.; Veroslavsky, G. Conchóstracos de la Formación Castellanos (Cretácico Inferior) de Uruguay (Cuenca de Santa Lucía). *Simp. Sobre Cretaceo Bras.* **1999**, *5*, 181–188. (In Spanish)

176. Vannier, J.; Thiéry, A.; Racheboeuf, P.R. Spinicaudatans and ostracods (Crustacea) from the Montceau Lagerstätte (Late Carboniferous, France): Morphology and palaeoenvironmental significance. *Palaeontology* **2003**, *46*, 999–1030. [CrossRef]
177. Gallego, O.F.; Zavattieri, A.M.; López–Arbarello, A. Conchóstracos y restos de peces de la localidad tipo de la Formación Río Mendoza (Triásico Medio), Provincia de Mendoza, Argentina. *Ameghiniana* **2004**, *41*, 289–301. (In Spanish)
178. Chunikhin, S.A. Triassic conchostracans from the Urengoy region of western Siberia. Proc. Tr. IV–Oy Mezhdunarodnoy Konferentsii. *Biniologiya Simmetrologiyai Sinergetika Yestestvennykh Nauk. Tyumen.* **2004**, *4*, 186–189. (In Russian)
179. Bishop, P.J. A Triassic conchostracan from near Murgon, SEQ. *Mem. Qld. Mus.* **2010**, *55*, 8–9.
180. Musacchio, E.A. Ostracodos de las superfamilias Cytheracea y Darwinulacea de la Formacion la Amarga (Cretacico Inferior) en la provincia de Neuquén República Argentina. *Ameghiniana* **1970**, *7*, 301–316.
181. Gallego, O.F.; Shen, Y.B. Revision of a conchostracan form from the La Amarga Formation (Lower Cretaceous), Neuquén Basin, Argentina. *Rev. Bras. Paleontol.* **2004**, *7*, 5–12. [CrossRef]
182. Tang, G.H.; Chen, N.S.; Xu, R.X.; Tu, F.L. Boundary division between Longtan Formation and Yanqiao Formation in northern Zhejiang Province. *J. Stratigr.* **1980**, *3*, 209–213. (In Chinese)
183. Wang, Y. New progress in the study of Jurassic and Cretaceous in the Greater Hinggan Mountains. *J. Stratigr.* **1985**, *9*, 203–209. (In Chinese)
184. Wu, Q.Q. Study on the subdivision and geological age of Huangmaqing Group of Nanjing district. *J. Chin. Acad. Geol. Sci.* **1980**, *1*, 65–90. (In Chinese with English Abstract)
185. Vallati, P. Conchostracos jurásicos de la Provincia de Chubut, Argentina. *Congr. Argent. Paleontol. Bioestratigrafía* **1986**, *4*, 29–38.
186. Zhang, W.; Dong, G.Y. Triassic System in Liaoning Province. *J. Stratigr.* **1982**, *6*, 20–32. (In Chinese)
187. Barth, G.; Kozur, H.W. A Latest Norian age for insect–bearing beds of the Fuchsberg and Langenberg near Seinstedt, northern foreland of the Harz Mountains (Lower Saxony, Germany). *Foss. Rec.* **2011**, *3*, 157–165.
188. Gallego, O.F. A new crustacean clam shrimp (Spinicaudata: Eosestheriidae) from the Upper Triassic of Argentina and its importance for 'conchostracan' taxonomy. *Alcheringa* **2010**, *34*, 179–195. [CrossRef]
189. Milner, S. Conchostraca from the Lower Jurassic Whitmore Point Member of the Moenave Formation, Johnson Farm, southwestern Utah. *Triassic–Jurass. Terr. Transit.* **2006**, *37*, 421.
190. Zheng, Y.J.; Su, F.; Chen, S.W.; Zhang, J.; Huang, X.; Gong, F.H. New discovery of fossils in the Lower Triassic Xingfuzhilu Formation, Bairin Right Banner, Inner Mongolia. *Geol. Bull. China* **2013**, *32*, 1423–1435. (In Chinese with English Abstract)
191. Depéret, C.; Mazeran, P. Les *Estheria* du Permian d'Autun. *Société D'histoire Nat. D'autun Bull.* **1912**, *25*, 165–172. (In French)
192. Wang, S.E. New Jurassic–Cretaceous conchostracans from Northern Hebei and Nei Mongol. *Acta Palaeontol. Sin.* **1984**, *23*, 726–729. (In Chinese)
193. Wang, S.E. The Jurassic and Cretaceous conchostracans from Sinkiang Uygur Autonomous Region, China. *Prof. Pap. Stratigr. Palaeontol.* **1985**, *12*, 1–21. (In Chinese with English Abstract)
194. Deng, S.H.; Yao, Y.M.; Ye, D.Q.; Chen, P.J.; Jin, F.; Zhang, Y.J.; Xu, K.; Zhao, Y.; Cheng, Y. *The Jurassic System of North China (I): A Stratigraphic Overview*; Petroleum Industry Press: Beijing, China, 2003; 339p. (In Chinese)
195. Liao, H.Y.; Shen, Y.B.; Huang, D.Y. Conchostracans of the Middle–Late Jurassic Daohugou and Linglongta beds in NE China. *Palaeoworld* **2017**, *26*, 317–330. [CrossRef]
196. Liao, H.Y.; Fang, S.H.; Feng, Z.; Gao, J.; Huang, D.Y. A new clam shrimp (Branchiopoda: Diplostraca: Spinicaudata) from the Upper Jurassic in the Jiyuan Basin, China and its biostratigraphic significance. *Zootaxa* **2023**, *5396*, 50–57. [CrossRef] [PubMed]
197. Liao, H.Y.; Cai, C.Y.; Feng, Z.; Huang, D.Y. Clam shrimp assemblage from Daohugou: A link correlating northern Hebei and western Liaoning. *Mesozoic* **2024**, *1*, 58–69. [CrossRef]

life

MDPI

Article

First Record of Aff. *Plesiosuchus* sp. (Mesosuchia, Metriorhynchidae) in the Kimmeridgian of Le Havre (Normandy, North-Western France): Biometry, Profile of Locomotion, and Paleobiological Consequences

Hua Stéphane

Odyssée Paléospace, Avenue Jean Moulin, F14640 Villers-sur-Mer, France; huasteph@aol.com

Abstract: This is the first record of the biggest Metriorhynchidae, aff. *Plesiosuchus* sp. in France. The remains consist of a partial vertebral column consisting of 11 vertebrae and an ischium fragment. A new method is proposed to evaluate the individual's size, which is estimated at 6.5 m. This method, unlike previous approaches, is based only on vertebrae and yields results that are congruent with those based on cranial remains. The state of preservation has allowed us to test the animal's 'profile of locomotion' to better interpret how it moved. Concerning other metriorhynchids, the record of *Dakosaurus* in France based only on teeth must be reassessed, and the genus *Torvoneustes*, if valid, has to be distinguished from *Plesiosuchus*.

Keywords: aff. *Plesiosuchus* sp.; Kimmeridgian; Normandy; vertebrae; biometry

1. Introduction

In January 1999, a block of still-embedded vertebrae and six free dorsal vertebrae was discovered by Franck Deschandol in a Kimmeridgian level at Ecqueville (Seine Maritime, north-western France). First described by Lepage et al. in 2008 [1], the mandible included in this paper was not found in association with the vertebrae. The material is kept at the Museum d'Histoire Naturelle du Havre under catalogue number MHNH 2007.5.531. The block has revealed five more nearly complete vertebrae, including a first sacral, as well as an ischium fragment from a marine crocodilian (Figures 1 and 2).

Such a discovery is not surprising in this area. However, what is surprising is the size of the vertebrae: they are in fact twice the size of other specimens found in the Lower Kimmeridgian of Octeville (e.g., *Metriorhynchus* cf. *hastifer* vertebrae, in [1], p. 55). As a group, the 11 vertebrae alone are 1.17 m long (Figure 1A).

The state of preservation of the animal, besides its taxonomy, raises questions about the size of the living animal. A new biometric approached is proposed to answer to this question.

Moreover, the state of preservation of such a column allows us to gain complete knowledge of the locomotion of this animal using the method of profile of locomotion.

The partial vertebral column of this specimen was displayed as part of the exhibition *Les Crocodiliens Fossiles de Normandie* during November and December 2008, after which it became part of the collections of the Muséum d'Histoire Naturelle of Le Havre.

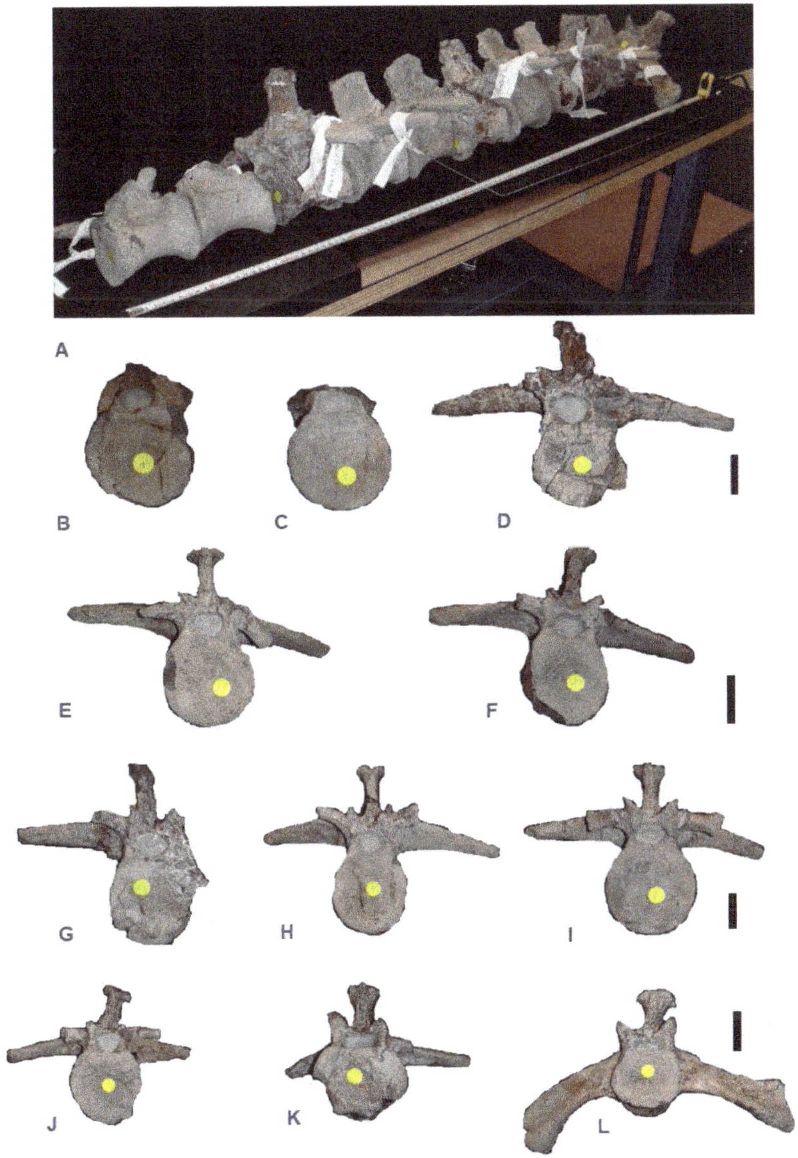

Figure 1. The vertebral column and all vertebrae in cranial view (scale bar = 5 cm) of aff. *Plesiosuchus* sp. (**A**) The vertebral column (MHNH 2007.5.530) shown in connection to give an idea of its impressive size (L = 117.9 cm); (**B**) 2007.5.531.9; (**C**) 2007.5.531.3; (**D**) 2007.5.531.13.1 (e); (**E**) 2007.5.531.6; (**F**) 2007.5.531.2; (**G**) 2007.5.531.13.1 (c); (**H**) 2007.5.531.1; (**I**) 2007.5.531.13.1; (**J**) 2007.5.531.13.1 (b); (**K**) 2007.5.531.13.1 (a); (**L**) 2007.5.531.13.1 (d).

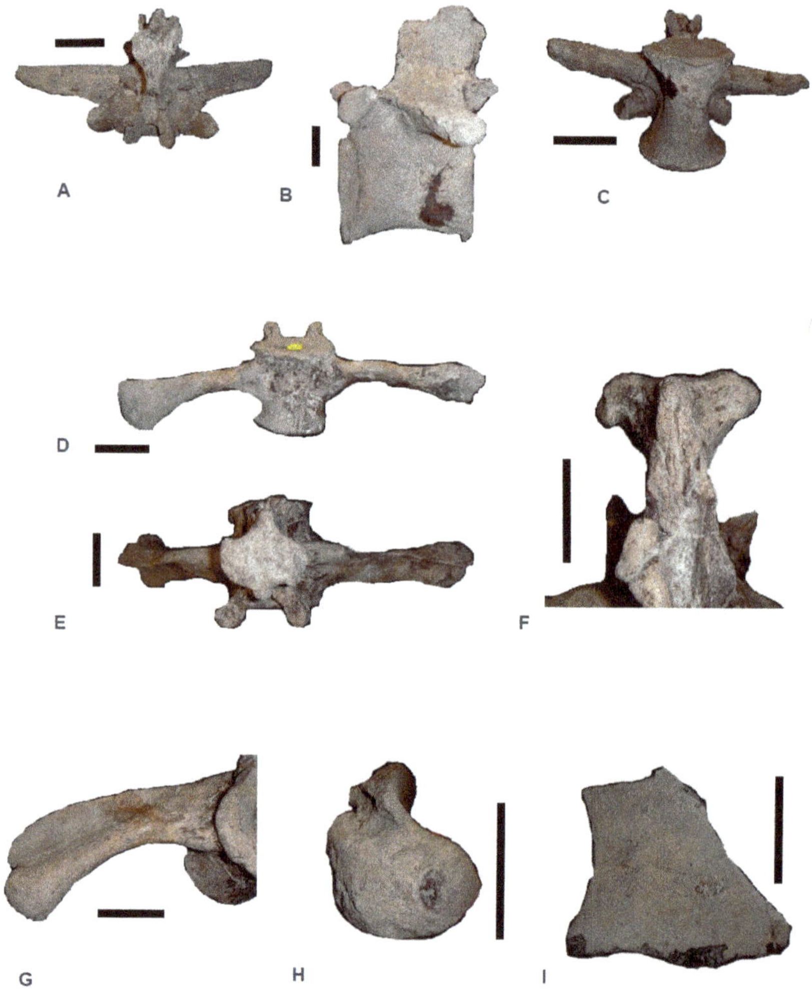

Figure 2. Different views of the dorsal and sacral vertebra described and the ischium. Dorsal vertebra 2007.5.531.13 in: (**A**) dorsal view; (**B**) left lateral view; (**C**) ventral view. Sacral vertebra 2007.5.531.13.1 (d): (**D**) ventral view; (**E**) dorsal view; (**F**) detail of the neural spine in posterior view showing size of muscular insertions and tuberosities. (**G**) Left sacral rib in anterior view and, (**H**) Left sacral rib in lateral view showing its "comma-shape" configuration, (**I**) fragment of left ischium in external view. Scale bars = 5 cm.

2. Geology

The location of the discovery is well identified thanks to Franck Deschandol, who discovered the specimen in January 1999 (with Gilles Lepage and Emmanuel Cacheleux) and subsequently donated it to the Museum of Le Havre (collection number 2007.5.530).

This crocodilian was found in a nodule in the middle part of the upper member of the Argiles d'Ecqueville formation, Pseudomutabilis Zone, strata XI.6 overlying the lumachelle beds, Upper Kimmeridgian [2]. Many vertebrates' remains have already been discovered embedded in limestone nodules with crystallized calcitic hearts.

3. Classification

The amphicoelous 'hourglass' vertebrae (Figure 1) are indicative of the grade mesosuchian. In such well-known marine levels, two mesosuchian families are possible: Teleosauridae and Metriorhynchidae. Isolated dorsal vertebrae are difficult to determine, but fortunately they have quite different sacral vertebrae.

In teleosaurids, such as the Kimmeridgian *Machimosaurus mosae*, sacral ribs are more or less horizontal [3], while in metriorhynchids, they face downwards (e.g., Andrews [4]), as in this specimen (Figure 1L). On the basis of this criterion, the specimen is clearly related to Metriorhynchidae.

Compared to Middle Jurassic Callovo–Oxfordian sites like Les Vaches Noires or the Oxford Clay Formation [5,6], the French Kimmeridgian is not very rich in terms of crocodilian specimens. This family is better represented in the British deposits of the Kimmeridge Clay Formation [7,8].

Concerning the genera, four have been identified in the Kimmeridgian: *Metriorhynchus* Meyer, 1830; *Dakosaurus* Quenstedt, 1856; *Torvoneustes* Andrade et al. [9]; and *Plesiosuchus* Owen, 1884.

Clearly, the robust and huge aspect of the vertebrae allows us to eliminate the genus *Metriorhynchus* and its slender vertebrae.

The taxonomic status of the genus *Torvoneustes* is far from clear: the type species *T. carpenteri* was initially considered a species of *Metriorhynchus*, then it was assigned to *Dakosaurus* in 2008 in Young et al. [7], to *Geosaurus* in 2009 by Young et al. [8], and finally to *Torvoneustes* in 2010 by Andrade and Young [9]. A new species, *T. coryphaeus*, was erected in 2013 [10] based essentially on cranial ornamentation (such criteria must be avoided due to intraspecific or ontogenetic variation).

Such a taxonomic attribution problem within the same team (cf. previously) is not aided by the fact that *Torvoneustes* and *Plesiosuchus* are sympatric and exhibit the same size range (could the specimen described in [10] as *Torvoneustes* in fact be a younger *Plesiosuchus*?). Most of the characteristics in the description of the type species by Young et al. [11] (p. 30 and Table 1) are essentially based on differences in tooth aspect. Even though tooth pattern is important for diagnosis, it must not be the main criterion for the diagnosis: intramandibular variation exists among thalattosuchians (e.g., *Machimosaurus mosae*, in plate 2 of Hua [3]) as well as ontogenetic variation.

The genus *Dakosaurus* is known from Europe [11] to South America [12,13], mainly based on cranial remains. Even though South American fauna are rich, post-cranial remains are rare. This genus is still poorly known in the Kimmeridgian of France, even though Foffa et al. [14] indicate that it was already present in the Callovian of Villers-sur-Mer without any clear references. All the French remains are isolated ziphodont teeth or poorly preserved remains ([11]), systematically reported as *Dakosaurus* sp. (e.g., [1], p. 96), even though ziphodonty is present not only in *Dakosaurus* but also in *Plesiosuchus* and *Torvoneustes*. Taking into account that in Young et al. [10], the only differentiating characters seem to be slight tooth ornamentation with few specimens known, the presence of *Dakosaurus* in France has to be reappraised. In fact, all French remains could perfectly well belong to *Plesiosuchus*. A similar problem is observed in blunt teleosaurid teeth: when found in the Callovian, they were systematically attributed to *Steneosaurus obstusidens*, whereas when they are found in the Kimmeridgian, they were attributed to *Machimosaurus hugii* (Hua et al. [15]) until a revision was carried out by Young, Hua et al. [16].

The Kimmeridgian genus *Plesiosuchus*, with *Plesiosuchus manselii* as the type species, is represented by a mandible and an incomplete skull. It was initially described by Owen in 1868 [17] as a pliosaur. In 1869, Eudes-Deslongchamps [18] demonstrated that the specimen was of metriorhynchid nature. Later authors confirmed this attribution, apart from Vignaud [12], who interpreted the specimen as *Machimosaurus mosae*. In 2012, Young et al. [11] clearly demonstrated the existence of *P. manselii* with an exhaustive description and diagnosis. In the absence of skull remains, as is the case here, all determinations must be considered tentative.

The size of the vertebrae and their shape, with their impressive tuberosities on the neural spine, are unknown for a specimen of genus *Metriorhynchus* of the same level [1], and for this reason it cannot be attributed to *Metriorhynchus*. *Plesiosuchus* vertebrae are unknown, but estimated size and stratigraphic level could make this genus as a good candidate.

A generic attribution would be too speculative, so this specimen is referred to as aff. *Plesiosuchus* sp.

The importance of this specimen is that it is the first relatively complete vertebral column referable to this genus and the first from French deposits.

As such, based on this discussion, the classification of this specimen is considered to be as follows:

Superorder Crocodylomorpha Hay, 1930 (*Sensu* Nesbitt, 2011) [19];
Order Mesoeucrocodylia Whestone and Whybrow, 1983 [20];
Suborder Thalattosuchia Fraas, 1901 [21];
Family Metriorhynchidae Fitzinger, 1843 [22];
Genus aff. *Plesiosuchus* Owen, 1884 (*sensu* Young et al. 2012 [11]).

4. Description

Only 3 out of 11 vertebrae will be described in detail to highlight their typical shape: a dorsal, a lumbar, and the sacral.

All measurements are provided in Table 1.

Table 1. Measurements of the 11 vertebrae of MNHN 2007.5.531. * Estimated measurement, total length, 117.9 cm.

Vertebrae	Anatomical Position	Inventory Number	Length of Centrum in Mid Ventral Line (cm)	Width of Anterior Face of Centrum (cm)	Height of Anterior Face of Centrum (cm)	Width of Posterior Face of Centrum (cm)	Height of Posterior Face of Centrum (cm)	Total Width	"Total Height (Neural Spine + Centrum)"	pre_Alpha	Post Alpha
Dorsal	6	2007.5.531.9	12.0	11.5	11.0	12.0	11	KO	KO	90	KO
D.	7	2007.5.531.3	12.5	12.0	11.5	10.0	10	KO	KO	KO	KO
D.	8	2007.5.531.13.1 (e)	10 *	8 *	8 *	10 *	10 *	34.5	KO	KO	KO
D.	9	2007.5.531.6	10.0	10.0	10.5	10.0	9	32	2105	90 *	75
D.	10	2007.5.531.2	10.5 *	10 *	10.0	11.0	10 *	27.5	22	85	55
D.	11	2007.5.531.13.1 (c)	11.0	10.2	10*	11 *	10	26 *	24	84	KO
D.	12	2007.5.531.1	11.0	9.9	10.3	10.0	10.8	32.2	23	82	60
D.	13	2007.5.531.13.1	10.0	11.0	11.0	11.0	10	28.5	22.5	92	60
D. (Lumbar 1)	14	2007.5.531.13.1 (b)	9.5	11.0	11.0	10.0	10	28 *	23	KO	90 *
D. (Lumbar 2)	15	2007.5.531.13.1 (a)	11 *	12.0	10 *	10 *	8 *	28 *	20	94 *	KO
Sacral 1	16	2007.5.531.13.1 (d)	10.4	11.5	8.0	8.1	8.4	37.6	18	90	84

The bauplan of these vertebrae do not differ from other Metriorhynchidae for the Callovian by Andrews [4] or for Oxfordian by Lepage et al. [1] (pp. 54–56) for the genus *Metriorhynchus*, apart from their huge and robust aspect.

The vertebra 2007.5.531.13 (Figures 1I and 2A–C) described here is considered to be a mid-dorsal vertebra. The centrum has an hourglass shape, like those of all mesosuchians, with the height of the anterior face being larger than the posterior one.

The articular ends of the centrum are moderately concave, with the vertical diameter being the greatest, and are slightly flattened at the level of the neural canal. Both faces of the centrum are slightly concave, and curiously for such an animal, no crenulations from muscular insertions are visible in the lateral view (Figure 2B). Unfortunately, no other such preserved metriorhynchid vertebral columns from the Kimmeridgian are known, but this needed to be described in case of a new discovery.

The most striking feature of these dorsal vertebrae is the robustness of the parapophysis process, which has a rounded section that is upward-oriented and is separated from the diapophysis by a distance of 3 cm. The articular surface of the parapophysis, with a diameter of nearly 1.5 cm (Figure 2A), indicates the robustness of the ribs.

The diaphyses are massive with the prezygapophyses included. The articular surface of the prezygapophyses is piriform, flat, and measures approximately 2 × 1 cm.

The diaphyses appear to be horizontal with a blade-like section in the anterior vertebrae and are more rounded and downward-oriented in the last dorsal vertebrae. In the dorsal view, the anterior border of the diaphysis process is convex and the posterior border is linear, finishing before the level of the anterior face of the centrum. The neural spines are developed in the last two-thirds of the centrum length and are relatively elevated with strong lateral tuberosity for muscular insertion, starting just after the posterior border of the parapophysis in the dorsal view. Such tuberosities are present and visible on skeletons of *Metriorhynchus*, but here it is very pronounced due to the size of the animal. Rectangular in lateral view, they overhang the posterior face of the centrum (Figure 2A).

The postzygapophyses are elongated, slightly convex, piriform in shape, and measure 2 × 1 cm.

In the dorsal area, it is interesting to highlight the difficulty of determining the number of lumbar vertebrae in metriorhynchids. Andrews [4] observed one or two lumbars; on the *Me.* cf. *superciliosus* described in Hua et al. [6], only one lumbar is identified with certainty; in the specimen described from the MNHN of Paris (MNHN 1908-6, pers. obs.), only one is visible; likewise, in this specimen, only one lumbar vertebra could be identified.

The vertebra 2007.5.531.13.1 (a) (Figure 1K) is interpreted as dorsal and more especially lumbar due to the smaller transverse process, which is downward-facing and not totally transversal like the others. Moreover, the paraphysis is fused with the diapophysis with a regressed parapophysis head of less than one cm. The cross section of the parapophysis is no more rounded than in the other vertebrae. The prezygapophysis still appears strong, with an angle of nearly 94°, and is totally independent from the diapophysis. The neural spine is shorter than the height of the centrum, but always bears strong transverse tuberosities due to muscular insertions. Despite its less well-preserved state, the centrum appears more circular than in the dorsal vertebrae.

In their general appearance, no dorsal vertebrae differ significantly from those described for another species of contemporaneous thalattosuchians, the teleosaurid *Machimosaurus mosae* [3]. The only difference is in the muscular tuberosities on the neural spines, which are much more developed on this specimen, indicating a more mobile trunk than is the case for *Machimosaurus*. This result is not surprising, as the trunk of *M. mosae* must have been more rigid with two osteodermal shields, one ventral and one dorsal.

The first sacral vertebra, 2007.5.531.13.1 (d) (Figures 1L and 2D–H), is complete and is typical of metriorhynchids, with sacral ribs facing downward and not transversally (e.g., as in *M. mosae*, plate 3 in [5]). The centrum is similar to that of the previous lumbar vertebrae. It is more massive, nearly quadrangular, with a section that is more transverse than vertical (the reverse for the dorsals). The prezygapophyses are well developed, with a surface measuring more than 2 × 1 cm, and are separated from the sacral ribs. In section, the ribs are in a comma-shape configuration (Figure 2H): a ventral border convex with a thin dorsal border. The connection with the ilium is 4 × 4 cm, below the level of the centrum.

The sacral ribs are inserted on the full height of the centrum, constricting at the middle to expand vertically for the insertion of the ilium over a height of more than 5 cm.

As for the lumbar vertebra, the neural spine is shorter than the height of the centrum, still with very pronounced transverse muscular tuberosities that protrude almost 2 cm beyond the surface of the neural spine (Figure 2F). The postzygapophysis appears weak in comparison to the size of the prezygapophysis.

The ischium is partial (Figure 2I), consisting of just a portion of the distal blade. This bone is thin (5 mm), showing slight convexity, as in all other metriorhynchids [4].

5. Size Estimation

Based on the work of Young et al. in [23], it is possible to estimate the size of an individual according to the size of the skull and the femur.

In this case, this was not possible because only eleven articulated vertebrae are present to work on. To estimate the size of this metriorhynchid, three nearly complete skeletons with articulated vertebrae that are well-preserved and not too distorted were used, as follows.

Me. cf. *superciliosus* FBS2012.4.67.80 described in Hua et al. [6], Callovian of the Vaches Noires, well preserved without the skull, estimated size 2.9 m.

Me. cf. *superciliosus* MPV 2010.3.610 described in Le Mort et al. [5], Callovian of the Vaches Noires, well preserved with partial skull, estimated size 2.5 m.

Me. superciliosus MNHN1908.6 exposed in the gallery from the Callovian of Peterborough, slightly crushed but complete, estimated size 1.90 m.

The skeleton of the teleosaurid, *Machimosaurus mosae* BHN2R1100, from the Kimmeridgian of Boulogne sur Mer, which is complete and well preserved with an estimated size of 5 m, has been also used.

The first surprising finding is that all the *Me. superciliosus* specimens used here are smaller (no more than 2.9 m) than the specimens measured in the study of Young et al. [8] (3 m to 3.5 m), also based on complete skeletons (but crushed). Furthermore, in estimating the size of a mesosuchian, an additional difficulty is the size of the intervertebral discs. This is always difficult to estimate, but in some complete skeletons, like the *Geosaurus suevicus* in the frontispiece of Andrews [4], these discs seem not to be too thick.

Young et al. [23] based their estimations on the length of the skull and the femur.

The approach here is totally different and independent based only on the size of vertebrae. By this method, the results could be compared if they are congruent.

With this specimen, two criteria have been used: the first one with all of them and the second one only using the first sacral vertebra.

The estimations are based on two criteria, detailed in Table 2.

The calculation is based on a basic rule of three, with measurements taken directly on the specimen.

Criterion 1: Total length of the 10 last dorsals + first sacral vs. total length of the skeleton.

Criterion 2: Length of the first sacral (not crushed) vs. total length of the skeleton.

Firstly, here the size of the adult specimens varies between 1.9 m and 2.9 m, whereas for Young [18], they vary from 3 m to 3.5 m (on crushed fossils). On complete skeletons of adults of the same species, ± 1 m of variation for adults of the same species appears coherent with what could be seen in the wild for living crocodylians of the same species of adults of different age.

Secondly, results obtained through comparison with *Machimosaurus mosae* show that isometric rules are valid only per family (even close like teleosaurids): the results provided by *M. mosae* are too different to be taken into account. The results here underline it.

According to the calculation of criterion 1, a total size of 6.5 m appears close to the result of Young et al. [11] (with a length of 6.8 m) using other isometric rules (based on skull and femora), whereas here only vertebrae were used.

As already mentioned, ± 1 m could appear for adults of the same species of various age, so this result appears congruent with the result of Young et al. [23] and can be taken to indicate the size of this specimen. Moreover, this study shows that results obtained using only the length of the vertebrae could be useful for estimating the length of such crocodilians.

The calculation based only on criterion 2, restricted to the first sacral vertebra, seems congruent and could be useful to estimate size on the basis of only one vertebra, which has never been done before.

Nevertheless, the results obtained are compatible with the results of Young et al. [11] (measured on a totally different biometric basis, confirming both results (this paper and the study of Young et al. [23])). *Plesiosuchus* is considered the giant of the Metriorhynchidae. This is not so surprising, for generally speaking, the marine crocodilian, including its living relative—*C. porosus*, the saltwater crocodilian—is known to reach a length of 7 m [24].

Table 2. Calculation of the total length of the specimen Tl according to the two criterias (related Cvert or just the first sacral) with different close relative specimens.

	M. superciliosus	*M.* cf. *superciliosus*	*M.* cf. *superciliosus*	*Machimosaurus mosae*	Aff. *Plesiosuchus*
	MNHN 1908-6	MPV 2010.3.610	FBS 2012.4.67.80	BHN2R1100	MHNH 2007.5.531
Crit1.: Total length of the (10D+1S) = Cvert (in cm)	34.1	45	53.5	68.4	117.9
Crit2.: Length of the S1 = LS1 (in cm)	3	3.9	4.5	5.1	10.4
Total length of the specimen (in cm) = Tl	240	250	290	500	Tl
size Aff. *Plesiosuchus* (Tl in cm) according to Crit 1 = f(specimen)	656.9	655	639.1	861.8	
size Aff. *Plesiosuchus* (Tl in cm) according to crit 2 = f(specimen)	658.7	666.7	670.2	1019.6	

Abbreviations: BHN, Boulogne sur Mer Musée d'Histoire Naturelle; FBS, Fabrique des Savoirs d'Elbeuf; MNHN, Muséum d'Histoire Naturelle de Paris; MHNH, Muséum d'Histoire Naturelle du Havre; MPV, Musée Paleospace de Villers.

6. Profile of Locomotion

First used by Hua [25], the technique applied here is based on the systematic angle of aperture of the zygapophyses across the vertebral column. This method of measurement was developed in order to compare dominant movements according to anatomic position between various taxa. Another advantage of this measurement is that it is independent of the thickness of the intervertebral disc and the intervertebral articulation, especially among crocodilians.

This is an advantage because from a purely anatomical point of view, and regardless of the classification used, two totally different kinds of articulations have historically existed in crocodilians: the amphicoelous vertebrae of the Protosuchians and Mesosuchians [26] and the procoelous vertebrae of the Eusuchians, i.e., 'flat' intervertebral articulation on the one hand and 'ball and socket' on the other. The latter development must have greatly increased the vertebral flexibility for eusuchians, regardless of taxon (Buffetaut [26]).

An additional difficulty in interpreting the axial mobility of the metriorhynchids is the fact that they were the only family without dorsal osteoderms, whereas some contemporaneous teleosaurids had three osteodermal shields (one dorsal and two ventral ones).

Therefore, when choosing between intervertebral joints or osteodermal shields to estimate the body mobility of such animals, we have to be careful to compare similar anatomic configuration.

Molnar et al. [27] tried to estimate the flexibility of various crocodilian taxa, including modern crocodilians and thalattosuchians, based in part on very interesting experimental results on living specimens. Nevertheless, these results must be tempered by the ball-and-socket intervertebral joint of the Eusuchians versus the acoelous articulation of the fossil taxa studied in this paper.

The locomotion profile of the specimen (in orange, Figure 3) is like that of other Metriorhynchidae for the thoracic region, where sharp angles indicate dominant vertical movements, whereas modern crocodilians undulate laterally when swimming.

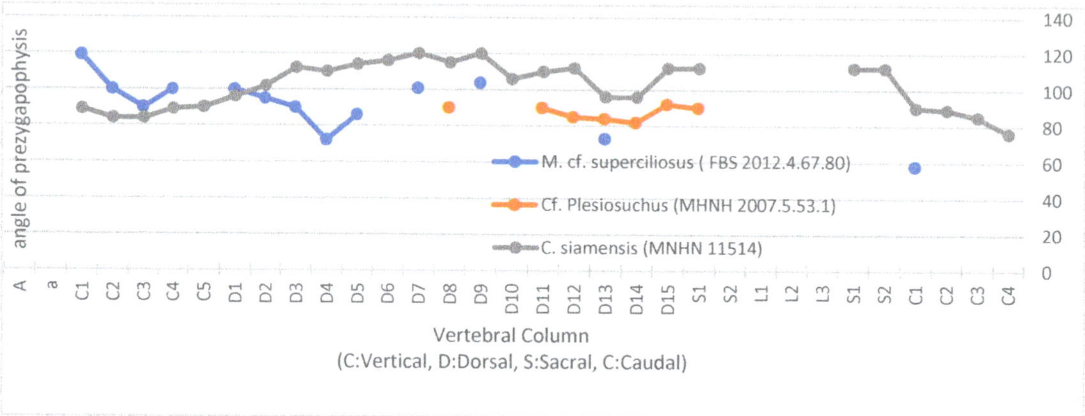

Figure 3. Aperture angle of prezygapophyses as a proxy for locomotion profile in a living eusuchian (*Crocodylus siamensis*) and a sub-complete *Metriorhynchus superciliosus* (Thalattosuchia, Metriorhynchidae) and aff. *Plesiosuchus*. Abbreviations: FBS, Fabrique des Savoirs d'Elbeuf; MNHN, Muséum d'Histoire Naturelle de Paris; MHNH, Muséum d'Histoire Naturelle du Havre; C, cervical; D, dorsal; S, sacral; C, caudal.

Interestingly, the angle of the aperture tends to exceed 90°, which is more than was observed for the other measured metriorhynchids by Hua [25]. Their trunk would thus have been flexible, allowing a greater degree of lateral movement than their close relatives like *Metriorhynchus*, while still maintaining greater rigidity for more efficient swimming than living crocodilians.

This anatomical polyvalency could indicate a behaviour different from the long-snouted and ichthyophageous *Me. superciliosus* noted by Hua [28]: the largest members of this family could have been more microphagous, as suggested by Molnar et al. [27]. For measurements made, all skeletons used were adults (neurocentral suture totally invisible), so ontogenetic variation can be discounted in the interpretation of the results.

This study underlines the fact that isolated vertebral column can be rich in information from the taxonomic point of view and also from a paleobiological point of view (biometric rules and locomotion profile).

7. Discussion: Thalattosuchians from the French Kimmeridgian

This partial skeleton, the first aff. *Plesiosuchus* to be confirmed in France, gives us a better understanding of thalattosuchian fauna during the Kimmeridgian.

The Kimmeridgian crocodilian fauna in France bear the largest thalattosuchian taxa: *Machimosaurus*, around 5–6 m in length for the Teleosauridae and *Plesiosuchus* and approximately 6.5 m in length for the Metriorhynchidae. More precisely, during the Callovian, both longirostrine (e.g., *Me. superciliosus*) and brevirostrine species (e.g., *Me. brachyrhynchus*)—Hua, 2008 in [1]—were about moderate size (around 2.5–3 m) and became more massive like *Plesiosuchus*, with double the size being observed for the first time during the Kimmeridgian in France.

It should be noted that this doubling phenomenon (increasing size combined with mesorostrine vs. longirostrine bipolarization) is the same for both members of the thalattosuchian group, Teleosauridae and Metriorhynchidae. The reason for this must be found with other coexisting epipelagic marine reptiles, like the other marine macropredators, the plesiosaurids, which were in regression during the Upper Jurassic, as described by Godefroit [29].

Besides size, the skull of such crocodilians converges in terms of shape towards the sauropterygian skull. This initially even misled Owen in his identification, especially because of the presence of a huge mandible without mandibular fenestrae, a unique feature among crocodilians.

The crocodilian skull is always a compromise between rigidity (to withstand torsion forces during the killing of prey) and lightness, phenomena underlined by Langston [30].

The closing of the mandibular fenestrae may have been a response to lightning caused by porosity of the cranial bones described by Hua and De Buffrénil [31] by increasing the rigidity of the skull. This modification would have led the skull shape to converge with that of the sauropterygians, explaining the initial mistake of Owen in 17].

The genus *Torvoneustes* must be reassessed to be differentiated with autapomorphic characters from the genus *Plesiosuchus*, given that for the moment, they cannot be safely distinguished.

These remains are from one of the largest metriorhynchids (approximatively 6.5 m in length) ever recorded in France. In addition, part of the axial skeleton survives in connection, the first such occurrence for the genus *Plesiosuchus.*

A new methodology has been used to estimate the animal's length, and the result seems to be corroborated by biometric measurements based on other criteria.

Funding: This research received no external funding.

Institutional Review Board Statement: Not applicable.

Informed Consent Statement: Not applicable.

Data Availability Statement: Enclosed in the study.

Acknowledgments: First, I would like to thank J. Parraga, curator of palaeontology at the Natural History Museum of Le Havre, for giving me access to this material. A big thanks also goes to the team of the museum for the preparation of the specimen and C. Ameury for their warm welcome. E. Buffetaut for his support and always precious advice. Y. Lepage provided invaluable help with the geological profile. Thanks also to Nor-Eddine Jalil, who facilitated access to a number of specimens in the Muséum National d'Histoire Naturelle. Thanks to my reviewers for their constructive advice!

Conflicts of Interest: The author declares no conflict of interest.

References

1. Lepage, Y.; Buffetaut, E.; Hua, S.; Martin, J.E.; Tabouelle, J. Catalogue descriptif, anatomique, géologique et historique des fossiles présentés à l'exposition "Les Crocodiliens fossiles de Normandie" (6 novembre–14 décembre 2008). *Bull. Soc. Geol. Normandie* **2008**, *95*, 5–152.
2. Samson, Y.; Lepage, G.; Hantzpergue, P.; Guyader, J.; Saint-Germain, S.M.M.; Baudin, F.; Bignot, G. Révision lithostratigraphique et biostratigraphique du Kimméridgien de la région havraise (Normandie, France). *Bull. Soc. Géol. Fr.* **1996**, *3*, 3–19.
3. Hua, S. The Crocodilian *Machimosaurus mosae* (Thalattosuchia, Teleosauridae) from the Kimmeridgian of the Boulonnais (Pas de Calais, France). *Palaeontogr. Abt. A Palaozool.* **1999**, *252*, 141–170. [CrossRef]
4. Andrews, C.W. *A Descriptive Catalogue of Marine Reptiles of the Oxford Clay;* Part Two; British Museum (Natural History): London, UK, 1913; 206p.
5. Le Mort, J.; Martin, J.E.; Picot, L.; Hua, S. First description of the most complete *Metriorhynchus* cf. *superciliosus* (Thalattosuchia) specimen from the Callovian of the Vaches-Noires cliffs (Normandy, France) and limitations in the classification of Metriorhynchidae. *Ann. Pal. Paris* **2022**, *108*, 102–139.
6. Hua, S.; Liston, J.; Tabouelle, J. The Diet of *Metriorhynchus* (Thalattosuchia, Metriorhynchidae): Additional Discoveries and Paleoecological Implications. *Foss. Stud.* **2024**, *2*, 66–76. [CrossRef]
7. Wilkinson, L.E.; Young, M.T.; Benton, M.T. A new metriorhynchid crocodile (Mesoeucrocodylia, Thalattosuchia) from the Kimmeridgian (Upper Jurassic) of Wiltshire, UK. *Paleontology* **2008**, *51*, 1307–1333. [CrossRef]
8. Young, M.T.; De Andrade, M.B. What is *Geosaurus*? Redescription of *Geosaurus giganteus* (Thalattosuchia: Metriorhynchidae) from the Upper Jurassic of Bayern, Germany. *Zool. J. Linn. Soc.* **2009**, *157*, 551–585. [CrossRef]
9. Andrade, M.B.; Young, M.T.; Desojo, J.; Brusatte, S.L. The evolution of extreme hypercarnivory in Metriorhynchidae (Mesoeucrocodylia: Thalattosuchia) based on evidence from microscopic denticle morphology. *J. Vert. Pal.* **2010**, *30*, 14–51. [CrossRef]
10. Young, M.T.; Andrade, M.B.; Brusatte, S.L.; Sakamoto, M.; Liston, J.J. The oldest known metriorhynchid super-predator: A new genus and species from the Middle Jurassic of England, with implications for serration and mandibular evolution in predacious clades. *J. Syst. Pal.* **2013**, *11*, 475–513. [CrossRef]

11. Young, M.T.; Brusatte, S.L.; Brandalise de Andrade, M.; Desojo, J.B.; Beatty, B.L.; Steel, L.; Fernandez, S.; Sakamoto, M.; Ruiz-Omenaca, J.I. The cranial osteology and feeding ecology of the Metriorhynchid Crocodylomorph genera *Dakosaurus* and *Plesiosuchus* from the late Jurassic of Europe. *PLoS ONE* **2012**, *7*, e44985. [CrossRef]
12. Vignaud, P. Les Thalattosuchia, Crocodiles Marins du Mésozoique: Systématique Phylogénétique, Paléoécologie, Biochronologie et Implications Paléogéographiques. Ph.D. Thesis, University of Poitiers, Poitiers, France, 1995.
13. Fernandez, M.S.; Herrerra, Y. Active airflow of the paranasal sinuses in extinct crocodyliforms: Evidence from a natural cast of the thalattosuchian *Dakosaurus andiniensis*. *Anat. Rec.* **2021**, *305*, 2604–9612. [CrossRef] [PubMed]
14. Foffa, D.; Young, M.T.; Brussate, S.L.; Graham, M.R.; Steel, L. A new metriorhynchid crocodylomorph from the Oxford Clay Formation (Middle Jurassic) of England, with implications for the origin and diversification of Geosaurini. *J. Syst. Palaeontol.* **2017**, *16*, 1123–1143. [CrossRef]
15. Hua, S.; Vignaud, P.; Pennetier, E.; Pennetier, G. Un squelette de *S. obtusidens* Andrews, 1909 dans le Callovien de Villers sur Mer (Calvados, France) et le problème de la définition des Teleosauridae à dents obtuse. *C. R. Acad. Sci. Paris* **1994**, *317*, 851–856.
16. Young, M.T.; Hua, S.; Steel, L.; Foffa, D.; Brusatte, S.L.; Thüring, S.; Mateus, O.; Ruiz-Omeñaca, J.I.; Havlik, P.; Lepage, Y. Revision of the Late Jurassic teleosaurid genus *Machimosaurus* (Crocodylomorpha, Thalattosuchia). *R. Soc. Open Sci.* **2014**, *1*, 140222. [CrossRef]
17. Owen, R. *Monography on the British Fossil Reptilian from the Kimmeridge Clay—Sauropterygia*; Paleontological Society: London, UK, 1868; Volume 3, p. 7.
18. Eudes-Deslongchamps, A. Prodrôme des Téléosauriens du Calvados. In *Notes paléontologiques 1867–1869*; LeBlanc-Hardel (Caen) & Savy: Paris, France, 1869; Volume 1, pp. 95–354.
19. Nesbitt, S.J. The Early Evolution of Archosaurs: Relationships and the Origin of Major Clades. *Bull. Amer. Mus. Nat. Hist.* **2011**, *2011*, 1–292. [CrossRef]
20. Whetstone, K.N.; Whybrow, P.J. A "Cursorial" Crocodilian from the Triassic of Lesotho (Basutoland). *South. Afr. Occas. Pap. Mus. Nat. Hist. Univ. Kans.* **1983**, *106*, 1–37.
21. Fraas, E. Die Meerkrokodile (*Thalattosuchia* n. g.) eine neue Sauriergruppe der Juraformation. *Jahresh. Des Ver. Für Vaterländische Naturkunde Württemberg* **1901**, *57*, 409–418.
22. Fitzinger, L.J.F.J. *Systema Reptilium*; Braumüller et Seidel: Wien, Austria, 1843; 106p.
23. Young, M.T.; Bell, M.A.; De Andrade, M.B.; Brusatte, S.L. Body size estimation and evolution in metriorhynchid crocodylomorphs: Implications for species diversification and niche partitioning. *Zool. J. Linn. Soc.* **2011**, *163*, 1199–1216. [CrossRef]
24. Wikipedia 2014—List of Largest Reptiles. Available online: https://en.wikipedia.org/w/index.php?title=List_of_largest_reptiles&oldid=1221630486 (accessed on 1 December 2024).
25. Hua, S. Locomotion in marine mesosuchians (Crocodylia): The contribution of the "locomotion profiles". *Neues Jahrb. Für Geol. Und Paläontologie-Abh.* **2003**, *227*, 139–152. [CrossRef]
26. Buffetaut, E. Radiation, Evolutive Paléoécologie et biogéographie des crocodiliens mésosuchiens. *Mémoires De La Société Géologique De Fr.* **1982**, *142*, 1–88.
27. Molnar, J.L.; Pierce, S.E.; Bhullar, B.A.S.; Turner, A.H.; Hutchinson, J.R. Morphological and functional changes in the vertebral column with increasing aquatic adaptation in crocodylomorphs. *R. Soc. Open Sci.* **2015**, *2*, 150439. [CrossRef]
28. Hua, S. Hydrodynamique et modalités d'allègement chez Metriorhynchus superciliosus (Crocodylia, Thalattosuchia): Implications paléoécologiques. *Neues Jahrb. Geol. Palaontol. Abh.* **1994**, *193*, 1–19. [CrossRef]
29. Godefroit, P. *Les Reptiles Marins du Toarcien (Jurassique Inférieur) Belgo-Luxembourgeois*; Service Géologique de Belgique: Brussel, Belgium, 1994; Volume 39, 98p.
30. Langston, W. The crocodilian skull in historical perspective. In *Biology of the Reptilia*; Gans, C., Bellairs, A., Parsons, T.S., Eds.; Academic Press: London, UK; New York, NY, USA, 1973; Volume 4, pp. 263–284.
31. Hua, S.; De Buffrenil, V. Histology of the Thalattosuchia as a clue of the interpretation of functional adaptations in the Thalattosuchian (Reptilia, Crocodylia). *J. Vert. Pal.* **1996**, *16*, 703–717. [CrossRef]

life MDPI

Article

Yuzhoua juvenilis: Another Angiosperm Seen in the Early Permian?

Xin Wang [1,*], Yinggang Lei [2] and Qiang Fu [1,3]

1 State Key Laboratory of Palaeobiology and Stratigraphy, Nanjing Institute of Geology and Palaeontology and CAS Center for Excellence in Life and Paleoenvironment, Chinese Academy of Sciences, Nanjing 210008, China
2 School of Materials Science and Engineering, Henan University of Science and Technology, Luoyang 471000, China; cchighkick@163.com
3 University of Chinese Academy of Sciences, Nanjing (UCASNJ), Nanjing 211135, China
* Correspondence: xinwang@nigpas.ac.cn

Abstract: "How old are angiosperms" is a frequently asked but still unanswered question. Although the underlying reason includes a lack of consensus on the criterion for fossil angiosperms, limited fossil finds, and other factors, the final answer to the question apparently lies in fossils, not wrangling among different schools. The currently mainstream idea in palaeobotany is that angiosperms cannot have existed earlier than the Early Cretaceous. This 64-year-old stereotype was recently iterated again in 2017. However, another hard-to-ignore fact is that this view is challenged by increasing pre-Cretaceous fossil evidence of angiosperms as well as molecular clock estimates. Here, we report a Permian angiosperm, *Yuzhoua* gen. nov. from Henan Province, China. This fossil plant has enclosed ovules, a defining feature idiosyncratic of angiosperms. In addition, a conspicuous style is seen on the top of the ovary, underscoring its distinction from known fossil seeds in gymnosperms. The combination of the Permian (Palaeozoic) age and these two unique features of *Yuzhoua* indicates that angiosperms first appeared much earlier than widely accepted, implying a much longer history of flowering plants. The occurrence of four specimens preserved in various states and unique morphology of *Yuzhoua* are beyond the expectations of all known theories on plant evolution, shedding new light on a previously unknown aspect of plant evolution in geological history.

Keywords: angiosperm; Permian; Palaeozoic; evolution; China; *Yuzhoua*

1. Introduction

Seventy years ago, Axelord argued that angiosperms may have existed in the Palaeozoic [1], and Stebbins stated that "the evolutionary line leading to the angiosperms entered a dark tunnel of ignorance at the end of the Paleozoic until the early Cretaceous" [2]. Their conclusions have two implications: (1) angiosperms first appeared in the Palaeozoic; (2) angiosperms, although present on Earth in the Palaeozoic, are missing from the fossil record of the early Mesozoic. Obviously, such a lacuna in our knowledge about angiosperms is unacceptable in botany, as it hinders us from a comprehensive understanding of plant evolution. This explains why there are so many botanists trying to decipher the origin and evolution of angiosperms and why there is an enormous amount of disagreement among botanists. Despite Axelrod and Stebbin's profiles and expertise in botany, their judgments are not coherent with the conclusion drawn by Scott et al. [3]. After scrutinizing all fossil evidence available then, Scott et al. rejected all pre-Cretaceous records of angiosperms in 1960 [3]. Since then, no widely accepted pre-Cretaceous angiosperms have been reported. Although there has been some research that suggests the existence of pre-Cretaceous angiosperms [4–14] and molecular dating has suggested the occurrence of angiosperms in the Jurassic or earlier ages [15,16], mainstream palaeobotanists insist on the ossified stereotype about angiosperm history. The latest reinforcement for such an idea was given by Herendeen et al. in 2017 [17]. The debate over angiosperm origin and history appears hard to conclude, probably because of (1) a lack of well-documented fossil angiosperms from the pre-Cretaceous, and (2) a lack of consensus on identification criteria for fossil angiosperms. Some botanists appeared to mix plesiomorph with apomorph, some proposed two different criteria for fossil angiosperms at the same time, while some others proposed a criterion that they themselves refused to adopt (for more details, see the discussion) [17–20].

Actually, it is not too hard to solve the problem of identification criteria for angiosperms. Going back to 1690, when the term "angiosperm" was coined by German botanist Paul Hermann, the term designated plants that have their seeds enclosed, in contrast to gymnosperms that have their seeds exposed to the exterior. A practical criterion for fossil angiosperms in palaeobotany is that the "ovules are completely enclosed in a carpel", according to which an Early Cretaceous angiosperm *Archaefructus liaoningensis* was identified and published in *Science* [21]. This criterion was later refined by Tomlinson and Takaso [22], who distinguished angiosperms from gymnosperms more strictly by the timing order of two developmental events: pollination and ovule enclosure. If ovules are enclosed before pollination, then the plant unexceptionally is an angiosperm [22]. Applying this criterion, various angiosperms have been identified in the Jurassic and earlier ages [11–14]. However, the records of early angiosperms are still sparse, not enough to topple the 'no-angiosperms-until-Cretaceous' stereotype. To figure out the truth of angiosperm origin and evolution, adopting Tomlinson and Takaso's criterion for angiosperms [22], here we pin down the angiospermous affinity of *Yuzhoua* from the Lower Shihhotse Formation (the Lower Permian) of Henan, China. The Permian age (>294.4 Ma) of *Yuzhoua* at least doubles the length of the widely accepted history of angiosperms, which was formerly thought to be only about 130 Ma. The unique morphology and early age of *Yuzhoua* reshape our understanding of early angiosperms and update our view on angiosperm evolution and history.

2. Materials and Methods

The specimens that we analyzed here were collected by YL and Chaofan Hou from an outcrop of the Lower Shihhotse Formation (Lower Permian) near Hongshanzui, Mojie town, Yuzhou city, Henan Province, China (113.17634° W, 34.15086° N), which is just outside the Cathayasian Palaeobotanical Geological Park in Henan Province, China. The palaeobotany of the region has been well studied long ago [23]. The fossil plants previously reported from this region include 307 species in 111 genera of Sphenophyllales, Equisetales, Filicales, Lepidodendrales, Noeggerathiales, Peltaspermales, Pteridospermophytes, Cycadales, Dicranohpyllopsida, Ginkgophyta, Cordaitopsida, Coniferopsida, Gigantopteridales, and plants *incertae sedis* (for detailed information, see the Supplementary File) [23]. This assemblage is typical of the Lower Shihhotse Formation, Lower Permian [24]. The age is the Asselian [23], and the radiometric age is 294.4–295 Ma [25].

The specimens were preserved as coalified compressions embedded in yellowish siltstones. The general morphologies of the specimens were photographed using a Nikon D300S digital camera (Nikon, Tokyo, Japan). Details of the specimens were examined and photographed using a Nikon SMZ1500 stereomicroscope equipped with a Digital Sight DS-Fi1 camera (Nikon, Tokyo, Japan). Details of better-preserved specimens were observed and photographed using a MAIA 3 TESCAN SEM (scanning electron microscope) (Tescan GmbH, Dortmund, Germany) at the Nanjing Institute of Geology and Paleontology (NIGPAS), Nanjing, China. All photographs were saved in TIFF/JPEG format and organized together for publication using Photoshop 7.0.

3. Results

Genus *Yuzhoua* Wang, Lei et Fu, gen. nov.

Diagnosis: Fruit asymmetrical, elongated, including a long stalk, an ovary, and a distal style. Stalk rigid, straight, gradually transitioning to ovary. Ovary in axil of an involucre. Involucre bearing two rows of long spines along its margins and bracketing the ovary. Ovary round triangular, enclosing three oval bodies of variable sizes clustered to one side of the ovary, with a layer of tissue covering its top. Style on top of ovary, slightly curved.

Remarks: The Yuzhou Flora has been well documented previously [23]. The previously documented fossils show no resemblance to the fossils documented here. Therefore, we propose a new genus, *Yuzhoua*, for these new specimens.

Three ovules of variable morphologies are enclosed in the ovary. No seed coat or similar structure is observed in all specimens, suggesting that the oval bodies in the ovaries are ovules rather than seeds (at least not mature seeds). The presence of three ovules inside an ovary confirms the occurrence of enclosed ovules in *Yuzhoua*, suggestive of angiospermous affinity. Therefore, we propose placing *Yuzhoua* in the angiosperm category despite its incredibly old age (>294.4 Ma).

Etymology: *Yuzhoua* refers to the fossil locality in Yuzhou, Henan, China.

Type species: *Yuzhoua juvenilis* gen. et sp. nov.

Type locality: Hongshanzui, Mojie Town, Yuzhou City, Henan, China (113.17634° W, 34.15086° N).

Horizon: the Lower Shihhotse Formation.

Age: the Asselian, Early Permian (>294.4 Ma ago).

Species *Yuzhoua juvenilis* Wang, Lei et Fu, sp. nov.

(Figures 1–4 and S1)

Species diagnosis: the same as that of the genus.

Description: The fossils documented here include six coalified compressions embedded in four blocks of yellowish siltstones (Figure S1a–f). Some of the specimens are partially covered in sediment (Figures 1a and S1a,b). Each fruit includes an elongated stalk, an ovary, and a distal style and is 23.5~40 mm long and 6.4–10 mm wide (Figures 1a, 2a, 3a and S1a–f). The stalk is rigid, straight, and becomes wider distally, gradually transitioning to the ovary; it is 15~23 mm long, 0.3~0.45 mm wide in its base, and up to 5 mm wide distally (Figures 1a, 2a and S1a–e). The stalk furcates into an involucre and a branch, with the ovary inserted in the axil of the involucre (Figures 1a and 2a,c,e). The involucre is 8.2~13 mm long, 0.4~1.8 mm thick, and bearing two rows of long parallel protective spines along its two margins, while the branch is short, bears no spines, and is 6.5~10.8 mm long and 0.9~1.6 mm thick, appressing the ovary and terminating just below the top of the ovary (Figures 1a,b, 2a,c,e and S1a,e,f). There are up to 10 protective spines visible on each side of the ovary, with the lengths of the visible portion varying from 1 to 7 mm (Figures 1a,b, 2a, 3a and 4b). The ovary is inverted, round, and triangular in shape, 6.8 mm long and 5.3 mm wide, with a layer of tissue covering its top and three ovules clustered to the ovary wall near the involucre (Figures 1a–c and S1a,c–f). Three ovules of variable forms (the central one is triangular, the upper and lower ones are distorted oval) are 2.1~4 mm long, 1.8~2.3 mm wide, oval, and appressing each other, each with an inconspicuous oval body in it (Figure 1b,c). There is a single coalified layer partially outlining the upper and lower ovules (Figure 1b,c), which are isolated from the ovary wall and covering layer of the ovary (Figure 1b,c). The covering layer of the ovary is almost uniform in thickness, 0.24~0.27 mm thick (Figures 1b,c, 2c,d and 4a,b). A distal style is attached on the covering layer, slightly curved, slightly tapering distally, 4.8~12 mm long, and 0.6~0.8 mm wide at the base, tapering into 0.4~0.55 mm wide distally (Figures 1a,b, 2a,b, 3a and 4a). The cellular details of the organ may be preserved sometimes (Figure 4c). The sketch and reconstructions of the organ are shown in Figure 5a–d.

Figure 1. *Yuzhoua juvenilis*, general morphology and details. PB205390a. (**a**) Holotype of *Yuzhoua*. The arrow marks the tip of the branch. Scale bar = 5 mm. (**b**) Detailed view of the top portion of the ovary with a conspicuous distal style (black arrow) and several protective spins (1~10) of variable lengths. The rectangle is detailed in (**c**). Scale bar = 1 mm. (**c**) Further detailed view of the rectangle in (**b**), showing three clustered ovules inside an ovary that is secluded on the top by a layer of tissue. Note the border between the ovule and the covering layer of the ovary (between arrows). The deployment of the ovules is sketched in the upper left inset drawing. Scale bar = 0.5 mm.

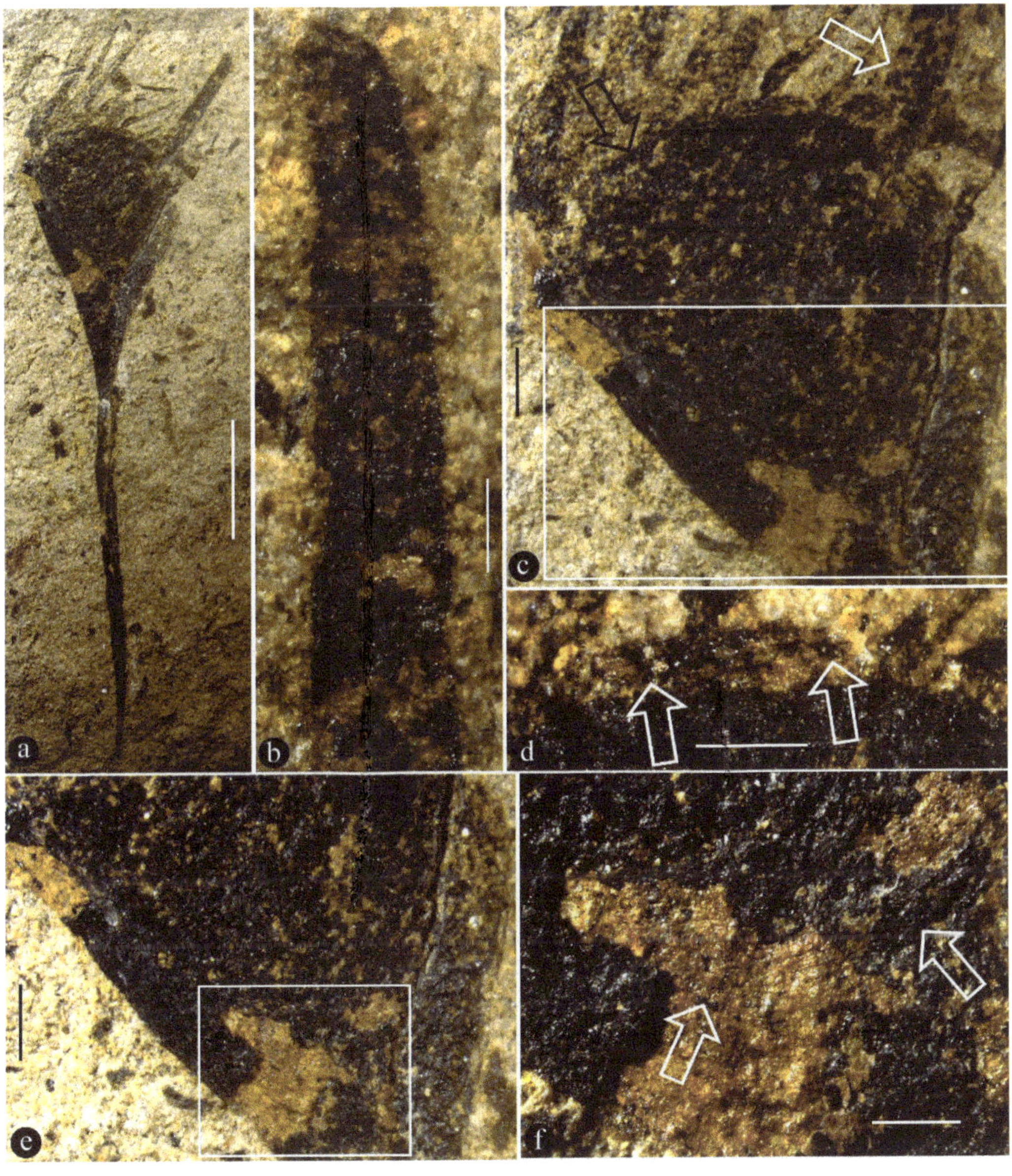

Figure 2. General morphology and the details of their lateral appendages. (**a**) General view. ParatPB205391. Scale bar = 5 mm. (**b**) Detailed view of the distal style. Scale bar = 0.5 mm. (**c**) Detailed view of the ovary, showing the distal style (white arrow) connected to a layer of tissue secluding the ovary (black arrow) and several protective spins. The rectangle is detailed in (**e**). Scale bar = 1 mm. (**d**) Detailed view showing the layer of tissue (white arrows) secluding the ovary. Scale bar = 1 mm. (**e**) Detailed view of the rectangle portion in (**c**). The rectangle is detailed in (**f**). Scale bar = 0.1 mm. (**f**) Detailed view of the rectangle in (**e**), showing the curving texture (arrows) of an inside ovule. Scale bar = 0.5 mm.

Figure 3. Another specimen of *Yuzhoua juvenilis* and its details. Paratype, 41HIV0181a. (**a**) General view. The rectangle is detailed in (**c**). Scale bar = 5 mm. (**b**) Detailed view of the upper portion in (**a**), showing slightly curving and parallel arrangement of the spines. Note the top of the ovary (arrows). Scale bar = 0.5 mm. (**c**) Detailed view of the rectangle in (**a**), showing curving parallel textures. Scale bar = 0.5 mm.

Holotype: PB205390 (Figures 1a–c, 4a–c and S1a,b).

Paratypes: PB205391, 41HIV0181, 41HIV0182 (Figures 2a–f, 3a–c and S1c–f).

Collectors: Yinggang Lei, Chaofan Hou.

Collecting date: 22 June 2024.

Depository: PB205390, PB205391 are deposited in the Nanjing Institute of Geology and Palaeontology, Nanjing, China; 41HIV0181, 41HIV0182 are deposited in the Henan Natural History Museum, Henan, China.

Etymology: *juvenilis* refers to the immaturity of the fossil organ.

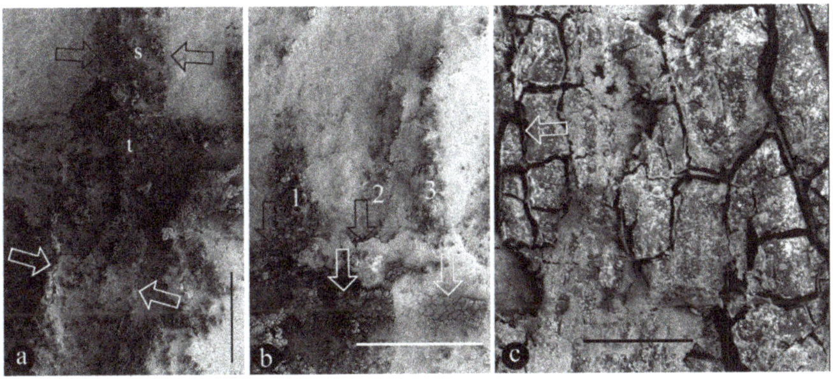

Figure 4. SEM details of *Yuzhoua juvenilis* shown in Figure 1a. PB205390a. (**a**) Style (s) physically connected to the top (t) of the ovary. Note the style (between black arrows) and ovary wall (between white arrows) of different thicknesses. Scale bar = 1 mm. (**b**) Detailed view of spines (1, 2, 3), top of ovary (black arrows), and ovule within the ovary (white arrows). Scale bar = 1 mm. (**c**) Cellular details of ovary wall. Note a thin cell wall (arrow) between cells. Scale bar = 0.1 mm.

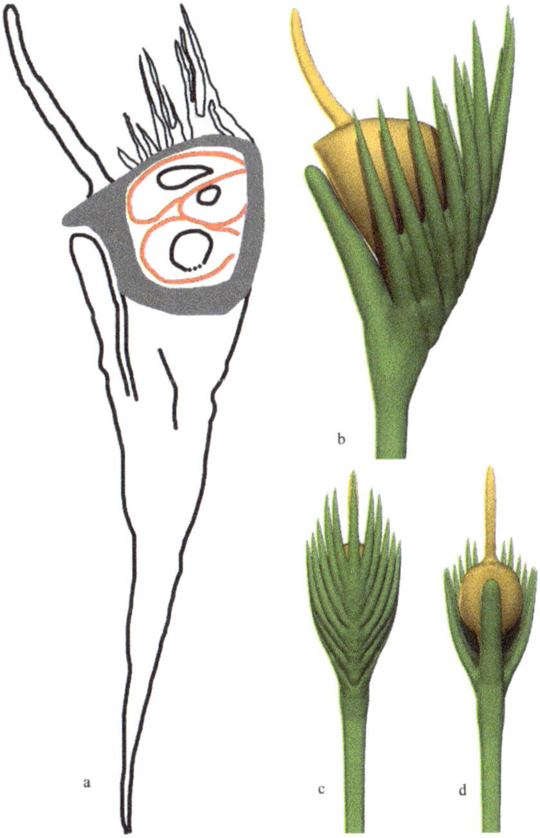

Figure 5. Sketch and reconstructions of *Yuzhoua* gen. et sp. nov. (**a**) Sketch of the specimen shown in Figure 1a. (**b–d**) Side, abaxial, and adaxial views of the organ.

Remarks: The general view of the organs shown in Figures 1a, 2a, and 3a excludes the possibility of them being a vegetative organ, leaving only one alternative for us: a reproductive organ. Among reproductive organs, there are three types: male, female, and hermaphroditic. Since both male and hermaphroditic organs require the presence of a male part, which encloses pollen grains that have more potential to be preserved in fossils than other tissues and organs, the lack of any trace of pollen grains in our light and SEM observations eliminates the possibility of *Yuzhoua* being male and hermaphroditic organs. This reasoning and inference are further strengthened by the dimensions (2.1~4 mm long, 1.8~2.3 mm wide) and inner position of the three oval bodies in Figure 1b,c. Therefore, we conclude that what we face in these specimens is a pistil.

The style of *Yuzhoua* may be otherwise interpreted as an extension of one of the branches, as it appears to be aligned with one side of the organ. However, this possibility can be easily eliminated: as shown in Figure 4a, both the ovary wall and style have parallel sides and uniform widths below and above the covering layer of the ovary, and it is clear that their widths are distinct, implying the style cannot be an extension of the ovary wall. Instead, the style is physically connected to the ovary top (Figures 1b, 2a,c and 3a).

Comparable to a mature fruit, *Yuzhoua juvenilis* gen. et sp. nov. appears immature as it lacks mature seeds that have a sclerenchymatic seed coat and tend to be preserved in fossils if present. The absence of seed in *Yuzhoua juvenilis* gen. et sp. nov. implies that what we see in Figure 1c are ovules (precursors of seeds). This explains why we used *juvenilis* as the specific epithet.

4. Discussion

Criterion for angiosperms: In 1690, the German botanist Paul Hermann (1646–1695) coined the term "angiosperms" for plants that enclose their seeds [26], in contrast to "gymnosperms" that have their seeds exposed to the exterior. Thus, "angiospermy" was the defining feature of angiosperms, at least initially so. If this criterion could be applied properly and consistently in palaeobotany, there would be or at least less controversy about the origin and evolution of angiosperms. However, there seems to be no consensus on a criterion for angiosperms in botanical circles; on one hand, those studying extant angiosperms focus on certain taxa (orders or families within angiosperms) and never worry about the scope of angiosperms [27]. They have no need to bother them about the criterion for angiosperms. On the other hand, those studying fossil angiosperms in the Cenozoic and Late Cretaceous are dealing with angiosperms that are already full-fledged and distinct from gymnosperms [28]. They have no need to worry about a criterion for angiosperms, either. The criterion for angiosperms matters only for botanists studying early angiosperms, as, theoretically, early angiosperms should demonstrate a smooth transition and show little difference from their contemporaneous gymnosperm peers. Ironically, these palaeobotanists seem to intentionally keep a tacit agreement on the criterion for angiosperms, as it is rarely declared in publications (except Sun et al. [21]). Various schools have their own unique criteria for angiosperms, especially when involved in discussions on early angiosperms and their origin [17–19]. For example, Sokoloff et al. emphasized pentamery, an apomorph of eudicots, as an angiosperm-specific feature [18], while Bateman preferred closed carpels and double fertilization as "a single reproductive syndrome in extant angiosperms" [19]. A fact hard to ignore is that Bateman was a member of Sokoloff et al. So, publishing these two conflicting papers in the same year raises a question for Bateman, "Which one of these two different criteria for angiosperms is your favorite?" Intriguingly, in the same paper, Herendeen et al. proposed several features as "unique angiosperm features", whereas they set five fossil angiosperms published by themselves as exemplar fossil angiosperms [17]. It is hard to believe that their own exemplar fossil angiosperms do not have all the features that they themselves proposed as "unique angiosperm features". Apparently, Herendeen et al. challenged the intelligence of the readers: readers would face a difficult choice about what to follow, their criterion or their examples. It appeared that Friis et al. [29] made a clear choice, "Neither" (as Friis et al. [29] ignored or refused to cite Herendeen et al.'s work even though it was frequently taken as a milestone in palaeobotany). Examining the author lists of these two papers, it is easy to find that Herendeen et al. equates to Herendeen plus Friis et al. Such an abnormal operation implies that Friis et al. are dissatisfied with their own published paper in *Nature Plants*. To avoid and reduce chaos on the criterion for fossil angiosperms, we prefer to return to the original meaning of "angiosperm" given by Hermann, which focused on "enclosed seeds".

A more practical criterion for fossil angiosperms was proposed and applied by Sun et al., which is the following: "The unique character of angiosperms is that the ovules are completely enclosed in a carpel" [21]. Using this criterion, Sun et al. published an Early Cretaceous angiosperm, *Archaefructus liaoningensis*, in *Science* [21]. This greatly contributed to the study of early angiosperms. Interestingly, Herendeen et al. also listed "carpels enclosing one or several bitegmic ovules with two integuments" as one of the "unique angiosperm features" [17]. It is noteworthy that both groups of authors converged on "ovules enclosed" in "carpels" as an angiosperm feature.

According to the traditional theories of angiosperm evolution, carpels in apocarpous gynoecia of Magnoliales and Ranunculales were taken as basic units of a gynoecium in angiosperms [30–32]. Naturally, carpels have long been a focus of botanical studies in the past decades [33–38]. However, this focus of studies becomes blurry since *Amborella* forced *Magnolia* out of the center of the stage in the molecular age [39]. Magnoliales and Ranunculales are now treated as Mesangiosperms instead of basalmost angiosperms [40]. When all angiosperms are taken into consideration, it is easy to find that *bona fide* carpels are restricted into limited apocarpous groups, and most angiosperms do not have typical carpels. Whether these angiosperms have carpels is an open question. Regardless of whether all angiosperms have carpels, an essential feature of angiosperms is still that the ovules are enclosed. To avoid unnecessary controversy, the present authors emphasize that ovules of angiosperms are enclosed in and by whatever structures.

The term "angiosperm" was coined in 1690, while the first usage of the term "carpel" was in 1835. This timing order indicates that Paul Hermann did not know of the term carpel. This explains why he did not specify which structure encloses ovules in angiosperms. So, in 1690, when he coined the term "angiosperm", he focused on the only distinction between angiosperms and gymnosperms, namely, angiospermy and gymnospermy. Apparently, although both Sun et al. [21] and Herendeen et al. [17] emphasized ovules enclosed "by carpels", "carpel-enclosing" as a character of angiosperms was appended to the original definition of "angiosperms" after botanists focused their studies on the assumed basalmost angiosperms with apocarpous gynoecia such Magnoliales and Ranunculales. Since these groups are not regarded as basalmost angiosperms anymore [41], whether carpels are still a feature required for all angiosperms becomes an open question. The present authors tend to reject the concept of "carpel" as it may be introduce unnecessary misunderstanding on angiosperms and may be a gratuitous extra requisite for angiosperms. This treatment is at least favored by the "noncarpellate" flower of a Jurassic angiosperm, *Nanjinganthus* [42–44].

The above treatment is consistent with the conclusion of Tomlinson and Takaso [22]. After a survey of pollination processes in seed plants, Tomlinson and Takaso distinguished angiosperms from gymnosperms by "ovules enclosed before pollination", and they did not mention which structure encloses ovules in angiosperms, either [22]. The drawback of this criterion is that it incorrectly places some angiosperms (for example, *Reseda*) as non-angiosperms, while its advantage is obvious: any plant meeting this criterion is unexceptionally an angiosperm. This explains why the present authors prefer to adopt this criterion for fossil angiosperms.

Actually, this criterion has long been applied in palaeobotany implicitly. In addition to the application of this criterion on *Archaefructus liaoningensis* [21], the same criterion has been applied in other studies on early angiosperms (for example, *Schmeissneria* [11,13], *Nanjinganthus* [42–44], and *Taiyuanostachya* [14], to name a few). *Schmeissneria* was initially recognized as an angiosperm based on an ovule impression on the inner-ovary wall and sealed ovary tip in a Chinese material, and its seeds expected inside the fruits were later confirmed by numerous German specimens [11,13]. *Nanjinganthus* has a couple of ovules enclosed inside the ovary, but shows no trace of carpels [42–44]. Considering its Early Jurassic age, it is hard to derive its gynoecium in a way other than the invagination of the floral axis. This explains the lack of *bona fide* carpels in *Nanjinganthus*. *Taiyuanostachya*, which is contemporaneous with *Primocycas* (Cycadales) [45], was recognized as an angiosperm because of not only its enclosed seed but also its enclosed ovule [14]. The co-occurrence of angiospermy and angio-ovuly in *Taiyuanostachya* [14] ensures its angiospermous affinity.

Alternative Interpretations: Seeds are frequently seen in the Yuzhou Flora, for example, *Acanthocarpus acerosus* Yang, *A. obovatus* Xie, *A. superjectus* Xie, *Cardiocarpus cordai* (Gein.) Gu et Zhi, *Carpolithus beaniata* (Zhang et Mo) Yang, *C. glansiformis* Mo, *C. mitriformis* Yang, *C. taxiformis* Stockm. et Math., *Dioonocarpus ovatus* Hu et Zhu, *Holcospermum linruense* Yang, *Rhabdocarpus oliveri* Kidston, *Samaropsis pachyderma* Yang, and *Zhongzhoucarpus deltatus* Yang, to name a few [23]. Considering the Permian age, it is necessary to eliminate the possibility of *Yuzhoua* being a kind of known or unknown seed. Among the above listed, *Acanthocarpus acerosus* Yang has a 5 mm long and 1.5 mm wide "sharp needle-like beak", *A. obovatus* Xie has a 3 mm long sharp beak, and *A. lagenarius* Halle has a 5 mm long and 2.5 mm wide "blunt conical snout" [23,46]. These seeds are the only ones known with an apical projection and are thus more or less comparable to *Yuzhoua* in the Yuzhou Flora. However, these alternatives can be easily distinguished from *Yuzhoua* by their lack of elongated stalk and protective spines. We will drop these alternatives hereafter.

Lepidocarpon is a noteworthy lycopod that has an unexceptionally large megasporangium, which are sometimes termed as a "seed". Its large size and lateral laminae [47] make it more or less comparable to *Yuzhoua*, but *Lepidocarpon* lacks any structures comparable to the distinctly elongated stalk and conspicuous style of *Yuzhoua*. Furthermore, *Lepidocarpon*, as a lateral part of a strobilus, is usually aggregated into a cone, unlike the sole individual *Yuzhoua*.

If the whole structure of *Yuzhoua* were a seed, then this seed would be spiny and long-stalked. Such a combination of characters is rarely seen in known seeds. Furthermore, the seed coat expected for a seed is fully missing in *Yuzhoua*. Finally, the occurrence of more than two oval bodies (Figure 1b,c) within a seed defies interpretations as it is so far unheard of for any seed yet. This makes the seed interpretation far-fetched and hard to accept. Therefore, we drop this alternative interpretation and prefer to interpret *Yuzhoua* as an ovary that has an apical style and two or three ovules enclosed inside.

Theoretical implications: The Permian age and unique morphology of *Yuzhoua* fall beyond the expectations of all known angiosperm evolution theories. Supposedly, an early angiosperm from the Permian should have fruits comparable to those in basalmost angiosperms, such as *Amborella* (according to APG [41]) or *Magnolia* (according to previously prevailing theory [30,32]). However, the carpels of Magnoliales [48] and *Amborella* [49] are all distinct from *Yuzhoua* in morphology, especially in the lack of distal style. Such a discrepancy between theoretical expectations and actual fossil evidence casts serious doubt over these perspectives on angiosperm evolution. Apparently, theories that are only based on data from extant plants have little "predictive power" on the past angiosperms, which can only be reliably learned by studying fossil angiosperms.

A previously reported fossil, *Taiyuanostachya ovuilifera*, represents the debut of angiosperms in the Permian [14]. Although it may flourish in the Permian vegetation, representatives of angiosperms in the Permian are still too sparse or poorly understood to topple the decade-old 'no-angiosperm-until-Cretaceous' stereotype. To confirm the factual existence of angiosperms in the Permian, fossil plants with some features unique to angiosperms are required. *Yuzhoua* appears to meet such an expectation: distal style and enclosed ovules are two features never seen in any gymnosperms but frequently seen in angiosperms. Apparently, the discovery of *Yuzhoua* strengthens the implication given by *Taiyuanostachya ovuilifera* [14], namely, that angiosperms did occur in the Permian and in a diversified and flourished mode.

Yuzhoua is contemporaneous with *Primocycas chinensis* (the earliest confirmed fossil record of cycads [45,50]) and *Taiyuanostachya ovuilifera* (the earliest confirmed fossil record of angiosperms [14]), suggesting that angiosperms and cycads co-occurred in the Palaeozoic, a scenario envisioned [1,51,52] and suggested by some molecular studies in which angiosperms were taken as a sister group of all gymnosperms. Such a scenario of plant evolution has been downplayed in botany. *Yuzhoua* now seems to invite botanists to rejuvenate this scenario of plant evolution. All fossil plants contemporaneous with *Yuzhoua*, including *Taiyuanostachya ovuilifera* [14] and *Primocycas chinensis* [45], belong to the well-known Cathaysian Flora of the Lower Permian, which is characterized by the *Alethopteris–Emplectopteris–Tingia–Cathaysiopteris* assemblage that was widespread in Asia during the Early Permian [53]. The Permian age of these plants has been proven by various palaeobotanists (including leading palaeobotanists such as Dr. Xingxue Li and Dr. Zhiyan Zhou [53,54]). The occurrence of *Yuzhoua* as an angiosperm in the Yuzhou Flora, together with *Taiyuanostachya* [14], appears to start solving the puzzle raised by Stebbins decades ago [2]. Although fossil records of angiosperms are still sparse, we expect that future serious digging and studies on fossil plants in the Permian will bring up more exciting finds.

Although *Taiyuanostachya* and *Yuzhoua* both belong to angiosperms, these two taxa share few common features (including angio-ovuly), implying an unexpectedly great diversity of angiosperms in the Permian as well as an earlier origin of angiosperms (if monophyly is assumed for angiosperms). Meanwhile, it is noteworthy that both *Taiyuanostachya* and *Yuzhoua* are abundant in their local floras, as more than one of their reproductive organs have been seen on a single piece of specimen of limited dimensions (Figure S1b,c of present paper; Figure S1c of Wang and Fu [14])). These observations indicate that angiosperms were not only diversified but also abundant in Permian vegetation, at least in north China. Compared to *Taiyuanostachya*, which shows little resemblance to extant angiosperms, *Yuzhoua* bears a feature that is more frequently seen in extant angiosperms: its conspicuous distal style looks like a typical style frequently seen in extant angiosperms. The occurrence of a long style in *Yuzhoua* offers the circumventive implication that (1) its ovules are fully enclosed, (2) the competition track for pollen tube viability (hitherto only seen in angiosperms) already occurred in the Permian if a distal stigma is assumed for *Yuzhoua*. Again, the latter implication is at odds with the widely accepted assumption of primitive carpels, which are usually thought of as styleless [30], as seen in *Magnolia* or *Amborella*. The doubt cast over the current theories of angiosperm evolution by this inference cannot be overstated. Apparently, the credibility and truthfulness of many botanical theories should be reviewed carefully.

The new insights on angiosperm evolution brought by *Yuzhoua* are in line with the fossil records of insects from the Permian [55–57]. Pollination in angiosperms is frequently related to insects. Some angiosperms even establish species-to-species coupling relationships with certain species of insects. Although this kind of ecological relationship has been well-studied in extant ecosystems, the history of this ecological tie has been a focus of controversy among scientists. Insect pollen transportation has been suggested to have existed as early as the Permian [55,56]. The earliest record of insects with a long proboscis (a feature of insect pollinators [57]) suggests that insect pollinators had occurred by the Early Permian (Kungurian, ~276 Ma) [56,57]. Although these fossil insects were thought unrelated to angiosperms [55–57], this thought apparently should be reviewed while keeping our new fossil records of Permian angiosperms in mind. Such reviewing should also take the latest phylogenomic study in count, which proposed a much earlier origin of angiosperms than before [16]. It appears that data from different fields are converging to the same conclusion, making it promising to draw a more accurate picture of angiosperm evolution.

Gigantopterids with typical angiosperm venation are especially abundant in the Yuzhou Flora [23]. Although they are not found physically connected with *Yuzhoua*, it would pay doing further research to elucidate the features of reproductive organs of Gigantopterids, vegetative organs of *Yuzhoua*, and contemporaneous taxa. These works will be more intriguing when the occurrence of vessel elements (an anatomical feature otherwise almost restricted to angiosperms) in Gigantopterids [58] is taken into consideration.

5. Conclusions

Yuzhoua is a new fossil plant found in the well-studied Yuzhou Flora. Although previous works have proven the Early Permian age of the flora, our study on *Yuzhoua* reveals a previously unknown feature: enclosed ovules. Enclosed ovules are a characteristic restricted to angiosperms; thus, we propose to place *Yuzhoua* in angiosperms. This will be surprising to many since angiosperms were thought to be absent in any pre-Cretaceous age. Together with previous records of early angiosperms (especially *Taiyuanostachya* from the Early Permian), fossil insects, and molecular estimates, the discovery of *Yuzhoua* pins down the occurrence of angiosperms in the Permian. Now, the Permian appears to be a new arena for angiosperm studies, as new discoveries have broken the bottleneck in research and opened a new field for palaeobotanists who have been discouraged from thinking of pre-Cretaceous angiosperms. The contemporaneous occurrences of cycads and angiosperms in the Permian (Palaeozoic) draw a new picture of plant evolution, requiring us to refresh our thinking on the evolution of angiosperms as well as of seed plants in general.

Supplementary Materials: The following supporting information can be downloaded at: https://www.mdpi.com/article/10.3390/life15020286/s1, Figure S1. Six fruits of *Yuzhoua* gen. et sp. nov. (**a**) Holotype, one of two fruits on the same specimen is shown in (b). It is still partially covered by the sediments. PB205390a. Scale bar = 5 mm. (**b**) Two fruits on the same specimen. PB205390. Scale bar = 10 mm. (**c**) At least two fruits on this specimen. Paratype, 41HIV0181. Scale bar = 10 mm. (**d**) One of two fruits in Figure S1c. 41HIV0181a. Scale bar = 5 mm. (**e**) Another fruit. Paratype, PB205391. Scale bar = 5 mm. (**f**) Another fruit. Paratype, 41HIV0182. Scale bar = 5 mm.

Author Contributions: X.W. designed the research project and drafted the manuscript. Y.L. collected the specimens. Q.F. analyzed the data. All authors have analyzed the data, modified, finalized and approved the manuscript. All authors have read and agreed to the published version of the manuscript.

Funding: This research is supported by the National Key Research and Development Program of China (2024YFF0807601) and the National Natural Science Foundation of China (42288201).

Institutional Review Board Statement: Not applicable.

Informed Consent Statement: Not applicable.

Data Availability Statement: All data are reported in this paper.

Acknowledgments: We thank Jingjing Tang for helping with photography, Yan Fang for his help with photography using MAIA3 TESCAN SEM, and Chao Tan for his help with the reconstruction. We appreciate the kind help on English from Mike Pole at the Queensland Herbarium, Australia.

Conflicts of Interest: The authors declare no conflicts of interest.

References

1. Axelord, D. A theory of angiosperm evolution. *Evolution* **1952**, *6*, 29–60. [CrossRef]
2. Stebbins, G.L. *Flowering Plants: Evolution Above Species Level*; Harvard University Press: Cambridge, UK, 1974.
3. Scott, R.A.; Barghoorn, E.S.; Leopold, E.B. How old are the angiosperms? *Am. J. Sci.* **1960**, *258-A*, 284–299.
4. Cornet, B. The leaf venation and reproductive structures of a late Triassic angiosperm, *Sanmiguelia lewisii*. *Evol. Theory* **1986**, *7*, 231–308.
5. Cornet, B. Late Triassic angiosperm-like pollen from the Richmond rift basin of Virginia, USA. *Paläontographica B* **1989**, *213*, 37–87.
6. Cornet, B. The reproductive morphology and biology of *Sanmiguelia lewisii*, and its bearing on angiosperm evolution in the late Triassic. *Evol. Trends Plants* **1989**, *3*, 25–51.
7. Cornet, B. Dicot-like leaf and flowers from the Late Triassic tropical Newark Supergroup rift zone, U.S.A. *Mod. Biol.* **1993**, *19*, 81–99.
8. Cornet, B.; Habib, D. Angiosperm-like pollen from the ammonite-dated Oxfordian (Upper Jurassic) of France. *Rev. Palaeobot. Palynol.* **1992**, *71*, 269–294. [CrossRef]
9. Hochuli, P.A.; Feist-Burkhardt, S. A boreal early cradle of angiosperms? angiosperm-like pollen from the Middle Triassic of the Barents Sea (Norway). *J. Micropalaeontol.* **2004**, *23*, 97–104. [CrossRef]
10. Hochuli, P.A.; Feist-Burkhardt, S. Angiosperm-like pollen and *Afropollis* from the Middle Triassic (Anisian) of the Germanic Basin (Northern Switzerland). *Front. Plant Sci.* **2013**, *4*, 344. [CrossRef]
11. Wang, X. *Schmeissneria*: An angiosperm from the Early Jurassic. *J. Syst. Evol.* **2010**, *48*, 326–335. [CrossRef]
12. Wang, X. *The Dawn Angiosperms*, 2nd ed.; Springer: Cham, Switzerland, 2018; p. 407.
13. Wang, X.; Duan, S.; Geng, B.; Cui, J.; Yang, Y. *Schmeissneria*: A missing link to angiosperms? *BMC Evol. Biol.* **2007**, *7*, 14. [CrossRef] [PubMed]
14. Wang, X.; Fu, Q. *Taiyuanostachya*: An abominable angiosperm from the Early Permian of China. *J. Biotechnol. Biomed.* **2023**, *6*, 371–379. [CrossRef]
15. Li, H.-T.; Yi, T.-S.; Gao, L.-M.; Ma, P.-F.; Zhang, T.; Yang, J.-B.; Gitzendanner, M.A.; Fritsch, P.W.; Cai, J.; Luo, Y.; et al. Origin of angiosperms and the puzzle of the Jurassic gap. *Nat. Plants* **2019**, *5*, 461–470. [CrossRef] [PubMed]
16. Zuntini, A.R.; Carruthers, T.; Maurin, O.; Bailey, P.C.; Leempoel, K.; Brewer, G.E.; Epitawalage, N.; Françoso, E.; Gallego-Paramo, B.; McGinnie, C.; et al. Phylogenomics and the rise of the angiosperms. *Nature* **2024**, *629*, 843–850. [CrossRef]
17. Herendeen, P.S.; Friis, E.M.; Pedersen, K.R.; Crane, P.R. Palaeobotanical redux: Revisiting the age of the angiosperms. *Nat. Plants* **2017**, *3*, 17015. [CrossRef]
18. Sokoloff, D.D.; Remizowa, M.V.; El, E.S.; Rudall, P.J.; Bateman, R.M. Supposed Jurassic angiosperms lack pentamery, an important angiosperm-specific feature. *New Phytol.* **2020**, *228*, 420–426. [CrossRef]
19. Bateman, R.M. Hunting the snark: The flawed search for mythical Jurassic angiosperms. *J. Exp. Bot.* **2020**, *71*, 22–35. [CrossRef]
20. Wang, X. Criterion is a touchstone in study of early angiosperms. *Open J. Plant Sci.* **2021**, *6*, 91–93. [CrossRef]
21. Sun, G.; Dilcher, D.L.; Zheng, S.; Zhou, Z. In search of the first flower: A Jurassic angiosperm, *Archaefructus*, from Northeast China. *Science* **1998**, *282*, 1692–1695.
22. Tomlinson, P.B.; Takaso, T. Seed cone structure in conifers in relation to development and pollination: A biological approach. *Can. J. Bot.* **2002**, *80*, 1250–1273. [CrossRef]
23. Yang, G. *The Permian Cathaysian Flora in Western Henan Province, China–Yuzhou Flora*; Geological Publishing House: Beijing, China, 2006.
24. Li, X.-X.; Yao, Z.-Q. Current studies of gigantopterids. *Palaeontol. Cathayana* **1983**, *1*, 319–326.
25. Wu, Q.; Ramezani, J.; Zhang, H.; Wang, J.; Zeng, F.; Zhang, Y.; Liu, F.; Chen, J.; Cai, Y.; Hou, Z.; et al. High-precision U-Pb age constraints on the Permian floral turnovers, paleoclimate change, and tectonics of the North China block. *Geology* **2021**, *49*, 677–681. [CrossRef]
26. Harper, D.R. The Online Etymology Dictionary. Available online: https://www.etymonline.com/search?q=angiosperm (accessed on 8 February 2025).

27. Wu, Z.-Y.; Lu, A.-M.; Tang, Y.-C.; Chen, Z.-D.; Li, D.-Z. *The Families and Genera of Angiosperms in China, A Comprehensive Analysis*; Science Press: Beijing, China, 2003; p. 1209.
28. Guo, S.-X. Pliocene floras of western Sichuan. *Acta Palaeontol. Sin.* **1978**, *17*, 343–350.
29. Friis, E.M.; Crane, P.R.; Pedersen, K.R. *Hedyosmum*-like fossils in the Early Cretaceous diversification of angiosperms. *Int. J. Plant Sci.* **2019**, *180*, 232–239. [CrossRef]
30. Cronquist, A. *The evolution and Classification of Flowering Plants*, 2nd ed.; New York Botanical Garden: Bronx, NY, USA, 1988; p. 555.
31. Eames, A.J. *Morphology of the Angiosperms*; McGraw-Hill Book Company, Inc.: New York, NY, USA, 1961; p. 518.
32. Takhtajan, A. *Flowering Plants, Origin and Dispersal*; Oliver AND Boyd Ltd.: Edinburgh, UK, 1969; p. 310.
33. Endress, P.K. The flowers in extant basal angiosperms and inferences on ancestral flowers. *Int. J. Plant Sci.* **2001**, *162*, 1111–1140. [CrossRef]
34. Endress, P.K. The evolution of floral biology in basal angiosperms. *Philos. Trans. R. Soc. B: Biol. Sci.* **2010**, *365*, 411–421. [CrossRef]
35. Endress, P.K. Multicarpellate gynoecia in angiosperms: Occurrence, development, organization and architectural constraints. *Bot. J. Linn. Soc.* **2014**, *174*, 1–43. [CrossRef]
36. Endress, P.K. The morphological relationship between carpels and ovules in angiosperms: Pitfalls of morphological interpretation. *Bot. J. Linn. Soc.* **2019**, *189*, 201–227. [CrossRef]
37. Endress, P.K.; Doyle, J.A. Ancestral traits and specializations in the flowers of the basal grade of living angiosperms. *Taxon* **2015**, *64*, 1093–1116. [CrossRef]
38. Endress, P.K.; Igersheim, A. Gynoecium structure and evolution in basal angiosperms. *Int. J. Plant Sci.* **2000**, *161*, S211–S223. [CrossRef]
39. Qiu, Y.-L.; Lee, J.; Bernasconi-Quadroni, F.; Soltis, D.E.; Soltis, P.S.; Zanis, M.; Zimmer, E.A.; Chen, Z.; Savolainen, V.; Chase, M.W. The earliest angiosperms: Evidence from mitochondrial, plastid and nuclear genomes. *Nature* **1999**, *402*, 404–407. [CrossRef]
40. Wang, Y.; Li, Y.-D.; Wang, S.; Tihelka, E.; Engel, M.S.; Cai, C. Modeling compositional heterogeneity resolves deep phylogeny of flowering plants. *Plant Divers.* **2024**. [CrossRef]
41. APG. APG IV: An update of the Angiosperm Phylogeny Group classification for the orders and families of flowering plants. *Bot. J. Linn. Soc.* **2016**, *181*, 1–20. [CrossRef]
42. Fu, Q.; Diez, J.B.; Pole, M.; Garcia-Avila, M.; Liu, Z.-J.; Chu, H.; Hou, Y.; Yin, P.; Zhang, G.-Q.; Du, K.; et al. An unexpected noncarpellate epigynous flower from the Jurassic of China. *eLife* **2018**, *7*, e38827. [CrossRef]
43. Fu, Q.; Diez, J.B.; Pole, M.; García-Ávila, M.; Wang, X. Nanjinganthus is an angiosperm, isn't it? *China Geol.* **2020**, *3*, 359–361. [CrossRef]
44. Fu, Q.; Hou, Y.; Yin, P.; Diez, J.B.; Pole, M.; García-Ávila, M.; Wang, X. Micro-CT results exhibit ovules enclosed in the ovaries of *Nanjinganthus*. *Sci. Rep.* **2023**, *13*, 426. [CrossRef]
45. Zhu, J.N.; Du, X.M. A new cycad–*Primocycas chinensis* gen. et sp. nov. discovered from the Lower Permian in Shanxi, China and its significance. *Acta Bot. Sin.* **1981**, *23*, 401–404.
46. Halle, T.G. Palaeozoic plants from central Shansi. *Palaeontol. Sin.* **1927**, *2*, 1–316.
47. Taylor, T.N.; Taylor, E.L.; Krings, M. *Paleobotany: The Biology and Evolution of Fossil Plants*, 2nd ed.; Elsevier: Amsterdam, The Netherlands, 2009; p. 1230.
48. Canright, J.E. The comparative morphology and relationships of the Magnoliaceae. III. Carpels. *Am. J. Bot.* **1960**, *47*, 145–155. [CrossRef]
49. Buzgo, M.; Soltis, P.S.; Soltis, D.E. Floral developmental morphology of *Amborella trichopoda* (Amborellaceae). *Int. J. Plant Sci.* **2004**, *165*, 925–947. [CrossRef]
50. Miao, Y.; Liu, Z.J.; Wang, M.; Wang, X. Fossil and living cycads say "No more megasporophylls". *J. Morphol. Anat.* **2017**, *1*, 107.
51. Martin, W.; Gierl, A.; Saedler, H. Molecular evidence for pre-Cretaceous angiosperm origins. *Nature* **1989**, *339*, 46–48. [CrossRef]
52. Martin, W.; Gierl, A.; Saedler, H. Angiosperm origins. *Nature* **1989**, *342*, 132. [CrossRef]
53. Li, X.; Yao, Z. Carboniferous and Permian floral provinces in East Asia. In *Papers for the 9th International Congress of Carboniferous Stratigraphy and Geology*; Thomas, D., Pfefferkorn, H.W., Eds.; South Illinois University Press: Urbana, IL, USA, 1985; pp. 95–101.
54. Li, X.-X.; Zhou, Z.-Y.; Song, Z.-C.; Ouyang, S.; Cao, R.-J. Advances in the studies of paleobotany of China during the last decade. *Acta Palaeontol. Sin.* **1989**, *28*, 129–150.
55. Khramov, A.; Foraponova, T.; Węgierek, P. The earliest pollen-loaded insects from the Lower Permian of Russia. *Biol. Lett.* **2023**, *19*, 20220523. [CrossRef]
56. Khramov, A.V.; Bashkuev, A.S.; Lukashevich, E.D. The fossil record of long-oribscid nectarivorous insects. *Entomol. Rev.* **2020**, *100*, 737–844. [CrossRef]

57. Peña-Kairath, C.; Delclòs, X.; Álvarez-Parra, S.; Peñalver, E.; Engel, M.S.; Ollerton, J.; Peris, D. Insect pollination in deep time. *Trends Ecol. Evol.* **2023**, *38*, 749–759. [CrossRef]
58. Li, H.; Taylor, E.L.; Taylor, T.N. Permian vessel elements. *Science* **1996**, *271*, 188–189. [CrossRef]

life

MDPI

Article

Lowermost Carboniferous (Tournaisian) Miospore Assemblages from the July Field, Gulf of Suez, Egypt: Biostratigraphic and Palaeoenvironmental Implications

Ahmed Maher [1,*] and Jiří Bek [2]

[1] Department of Geology, Faculty of Science, Al-Azhar University, Assiut 71524, Egypt
[2] Laboratory of Palaeobiology and Palaeoecology, Institute of Geology, Academy of Sciences of the Czech Republic, Rozvojová 269, 165 00 Prague, Czech Republic; bek@gli.cas.cz
* Correspondence: ahmed.maher.21@shizuoka.ac.jp or maher4u20@azhar.edu.eg; Tel.: +20-1097790639

Abstract: The Nubia Sandstone in the Gulf of Suez, Egypt, is a well-known unclassified sediment. Palynology is considered the most effective tool for dealing with this problem. Miospore assemblages from the Lowermost Carboniferous (Tournaisian) have been discovered from the J62-86 and the J62-64 AST1 wells located in the July Field of the Gulf of Suez, Egypt. Spores are moderately to poorly preserved, suggesting a stratigraphical position within Lowermost Carboniferous ages. The studied sediments include poorly preserved conodont fragments and present significant identification challenges due to the drilling methodologies' complexities. Spore assemblage consists of 31 genera with 56 species. The dominant spores include zonate genera *Vallatisporites*, *Densosporites*, and *Archaeozonotriletes*, camerate genera *Grandispora*, *Geminospora*, apiculate genera *Apiculiretusispora*, and laevigate trilete genus *Punctatisporites* and megaspores of the *Lagenoisporites* type are recorded. Marine microphytoplankton including *Schizocystia bicornuta*, *Lophosphaeridium*, *Leiosphaerida*, and some filamentous green algae of unknown affinity are recorded. The dispersed spore assemblage is associated with carbonized plant fragments. The palynological data have effectively dated the lower intervals of the Nubia Sandstone from the Nubia "B", indicating a Lowermost Carboniferous (Tournaisian) age, i.e., *Vallatisporites vallatus–Retustriletes incohatus* palynozone (VI). The stratigraphic differentiation of the Nubia Sandstone is crucial for subsequent correlating subsurface wells in the Gulf of Suez within the context of geology and hydrocarbon exploration, particularly given the scarcity of other fossil groups.

Keywords: Carboniferous; Tournaisian; spores; Gondwana; Gulf of Suez; Egypt

1. Introduction

The megaflora and microflora of the Devonian–Carboniferous strata of the eastern part of Egypt have not yet been studied, especially in the eastern part of Egypt in the Gulf of Suez region. This is because of rare Devonian–Carboniferous outcrops and the challenges of finding fossil records in the Nubia Sandstone, representing unclassified sediments from the Cambrian to the Cretaceous age. There were no sufficient fossil records of this complex succession of sediments until today.

The fossil flora, including spores, is crucial for the stratigraphical position of sediments, particularly when fossil fauna and geological indicators are missing. Herein, the palynology is considered the only method for correlating the sedimentary rocks derived from the Nubia Sandstone.

https://doi.org/10.3390/life15060872

The understanding of the complicated subsurface stratigraphical relations of the Nubia Sandstone remains an argumentative subject despite its significance for the geological framework of North Africa and the Arabian Peninsula. Egypt's Devonian–Carboniferous stratigraphy is unknown, and no prominent outcrops have been identified. The subsurface well data are the only source of information for investigating this geological system. The evident infrequency of guide fossils within the Palaeozoic clastic sequences in Arabia and Northeastern Africa has significantly challenged their stratigraphic position [1].

1.1. The Complexity of the Geologic Setting of the Nubia Sandstone

The Nubia Sandstone (Figure 1) exhibits substantial potential for hydrocarbon exploration. The regional sedimentary sequence of the Nubia Sandstone covers various localities in Egypt and Northeastern Africa, extending throughout Libya, Sudan, Chad, Jordan, Saudi Arabia, and Yemen [2]. Several Palaeozoic outcrops have been documented in the Gulf of Suez region if they relate to the Nubia Sandstone. The Cambrian (Araba Formation) has been described from the Araba–Durba region in Southwestern Sinai and has been identified in the northern sectors of the Abu Gharadig Basin [3].

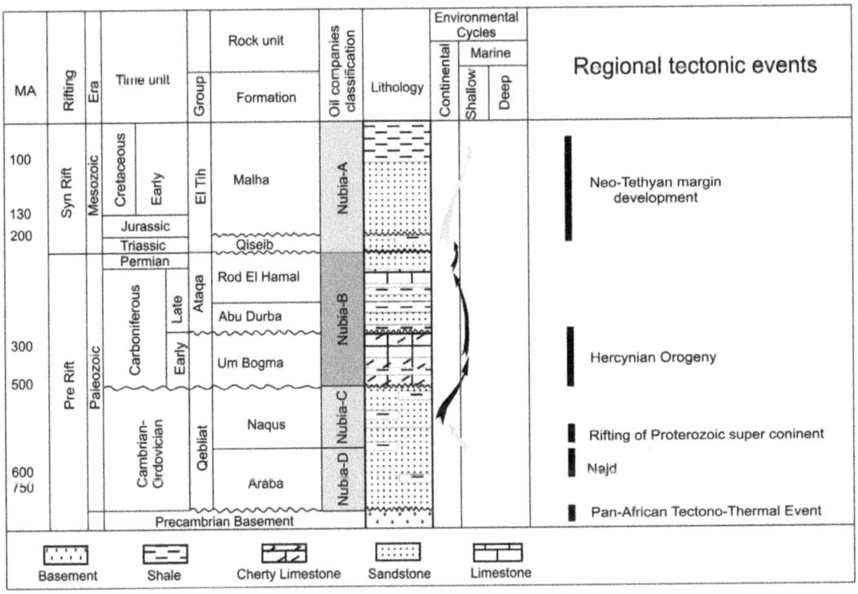

Figure 1. Generalized tectonic and lithostratigraphic column in the Gulf of Suez illustrating the Nubia Sandstone, Egypt (changed from [4–11]). The arrows refer to the environmental cycles.

The Ordovician–Silurian sequences are observed to overlie the Cambro-Ordovician Araba Formation on the western side of the Gulf of Suez [3]. Rare Carboniferous outcrops have been recognized in the Gulf of Suez region. The Early Carboniferous stratigraphic section in the Gulf of Suez Basin has been described from the Um Bogma Formation [12,13]. Although there is no sedimentary record of the Permian formation in Southern Egypt, the Permian might be represented in subsurface well data north of the Western Desert in the Siwa Basin; in the Gulf of Suez Basin, however, the Permian and Triassic formations developed as undifferentiated fluvial sandstone sequences attributed to the Quseib Formation [14]. No palynological data deal with this problem yet.

The Nubia Sandstone "D" member was correlated with the Um Bogma Formation and in the Eastern Sinai. The Nubia Sandstone "C" Member was equivalent to the Abu Thora

Formation of the Early to Late Carboniferous in the Um Bogma area, as well as partially correlating with the Rod El Hamal and Abu Darag formations of the Late Carboniferous in Wadi Araba on the western side of the Gulf of Suez. The Nubia "B" member was correlated with the Aheimer Formation of the Late Carboniferous [15]. The Abu Durba Formation of the Early Carboniferous, along with the Aheimer Formation from the Late Carboniferous, exhibits a conformable relationship with the Nubia "B" member; meanwhile, the uppermost section (designated as the Nubia "A" member) correlates with the Permo-Triassic Qiseib Formation located in Wadi Araba, situated on the western flank of the gulf [2,16]. The unconformable surface between the Nubia "C" and the Nubia "B" members shows the non-record of the Silurian and the Devonian ages in the eastern part of Egypt in the Gulf of Suez (Figure 1).

Deficiency in the chronological and stratigraphical dating of the Nubia Sandstone in the Gulf of Suez is known. The objective of this study is to define the undifferentiated sediments obtained from the Nubia Sandstone to gain an understanding of its formation processes and biostratigraphic sections. The study aims to identify the palynomorph assemblages in the Arabian Plate extending into northern Africa, specifically in Egypt.

1.2. Previous Work in the Middle East Areas and Egypt

The preceding Palaeozoic palynological studies of the Gulf of Suez and various regions in Egypt have focused on the Carboniferous palynology, as evidenced by the works of [17–24]. Recently, ref. [25] documented significant occurrences of the Carboniferous spore *Vestispora* (Wilson and Hoffmeister) Wilson and Venkatachala in the northwestern desert of Egypt, ref. [26] dated some strata to the Late Emsian to Early Givetian age through the identification of numerous palynomorphs, and [27] investigated spore assemblages believed to belong to the Devonian and Lower Carboniferous from the Kohla-2 well situated in the Western Desert of Egypt. There are insufficient records of the Devonian–Carboniferous boundary in Egypt.

1.3. Geologic Setting

The Gulf of Suez is located between the African and Arabian tectonic plates (Figure 2); it has been studied for oil and gas exploration [28]. The contemporary Gulf of Suez rift, along with the Red Sea oceanic basin and the Aqaba–Dead Sea transform systems, constitutes the Sinai triple junction (Figure 3), which originated during the northeastward displacement of the Arabian plate from the African plate [9]. The Gulf of Suez represents a large graben/half graben assemblage situated between two basement uplifts—the Sinai and the Eastern Desert mountain ranges [10]. The Palaeozoic–Lower Cretaceous Nubian Sandstone consists of a substantial sequence of clastic and thin carbonate sediments, reaching depths of up to 1200 m [11]. The subsurface strata in the Gulf of Suez are recognized as the Nubia Sandstones [11]. The July Field (Figure 3) is regarded as one of the principal oil fields in the Gulf of Suez, Egypt [29]. The Nubia Sandstone exhibits thick stratifications ranging from the Palaeozoic to the Mesozoic and is positioned unconformably on the top of the Precambrian basement complex; this formation is distributed throughout the Gulf of Suez with variable thicknesses, with these strata being exposed along the western and eastern margins of the Gulf of Suez [9]. The characteristics of the Nubia Sandstone are consistent and situated in proximity to the northern boundary of the African craton [30].

1.4. Stratigraphy of the Pre-Rift

The pre-rift stratigraphic sequences in the southern Gulf of Suez are characterized by a basal sandstone layer, which is subsequently overlain by an assemblage of interbedded shale, sandstone, and limestone (Figures 2 and 4). The basal section of the Nubia, in certain locations, is constituted of several tens of meters of interbedded red siltstone and sandstone that remains undated. The Palaeozoic succession of the Nubia Sandstone encompasses a series of siliciclastic and sedimentary intervals whose depositional characteristics are indicative of shallow marine to fluvio-deltaic environments [10,11].

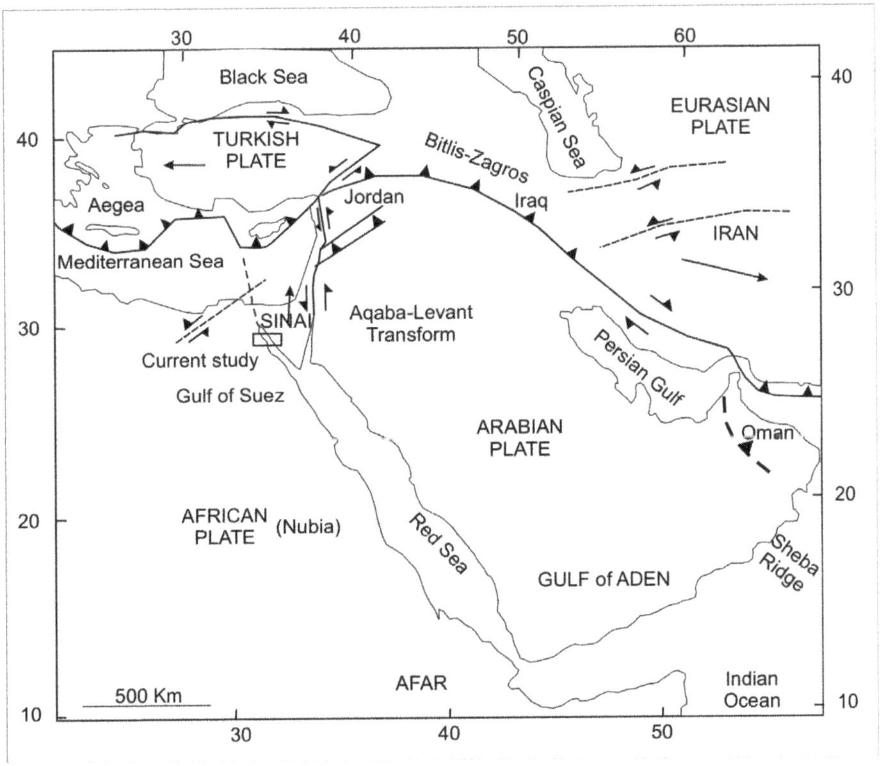

Figure 2. Simplified tectonic setting map showing the location of the Gulf of Suez, Middle East, with an illustration of plate motions (changed from [4,5]).

The Palaeozoic rocks exposed in the Um Bogma area in the Southwestern Sinai have been discussed in [23]. The correlation between the surface and subsurface Palaeozoic complex is illustrated in (Figure 1). The Ataqa Group (Figure 1) was used to describe different sections in Central Sinai boreholes and in the Eastern Desert [15]. The Ataqa Group is underlain by the Qebliat Group (Figure 1), and these units are dominantly clastic and were proposed to belong to the Cambro-Ordovician based on their trace fossil contents [23]. The interval between the Cambrian and Carboniferous strata (Figure 1) is poorly dated, and this unit was named Nubia "C" [15]. This unit of Nubia "C" is well represented in the July Field of the current study, and it was proposed to be of the Ordovician, Silurian, and Devonian age [15]. In Southwestern Egypt and Northwestern Sudan, the Devonian is not dated palaeontologically. However, part of the strata overlying the Silurian beds at Gebel Tageru, east and southeast of Gebel Kissu, and at Gebel Uweinat (Figure 4) were proposed to be of the Devonian age [30], while in northwestern Egypt, there were some palynomorphs reported to be of Late Emsian to Early Givetian age [26].

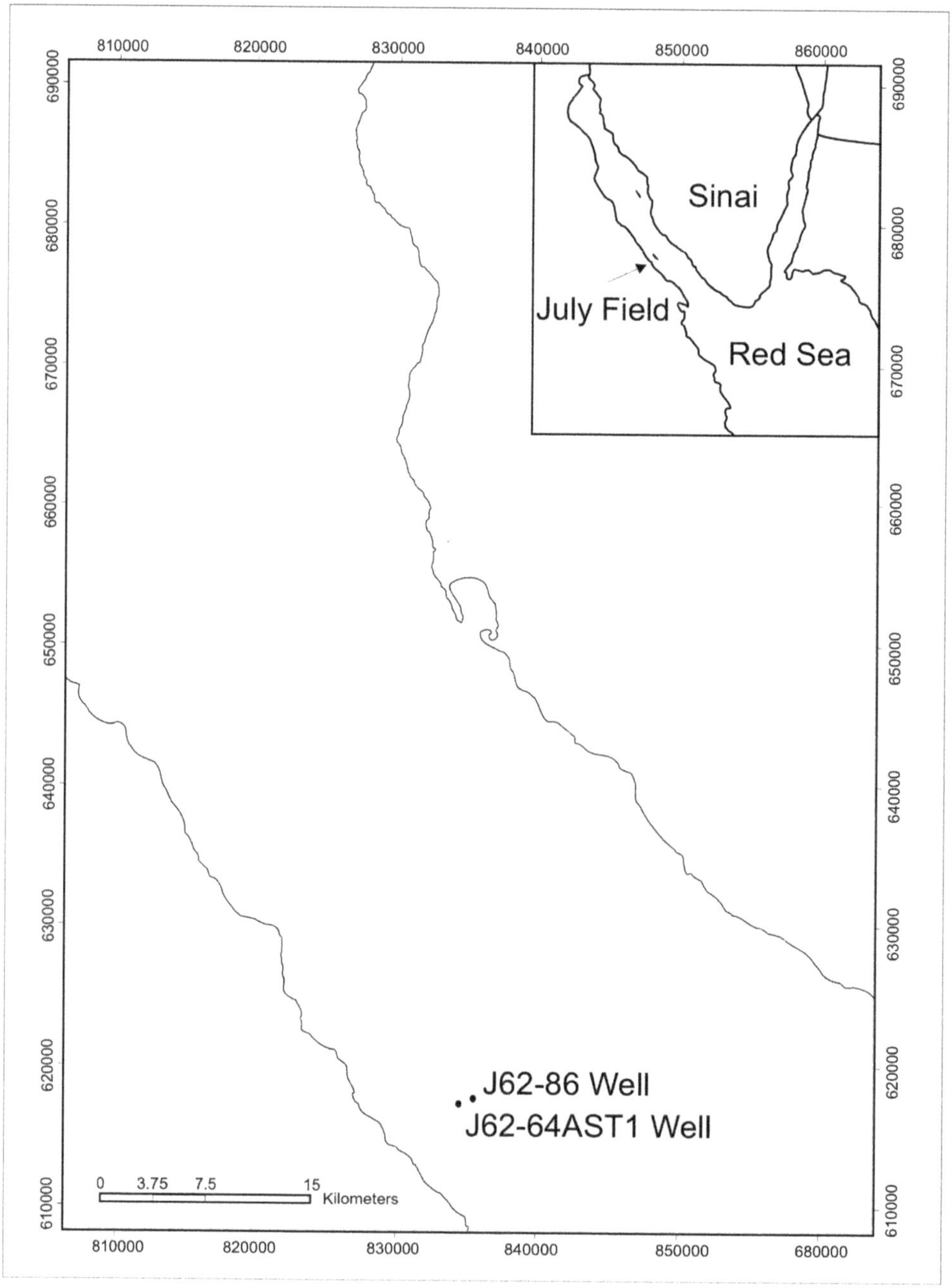

Figure 3. Location map in the July Field showing the J62-86 well and the J62-64 AST1 well [31,32]. The arrow refers to the July Field.

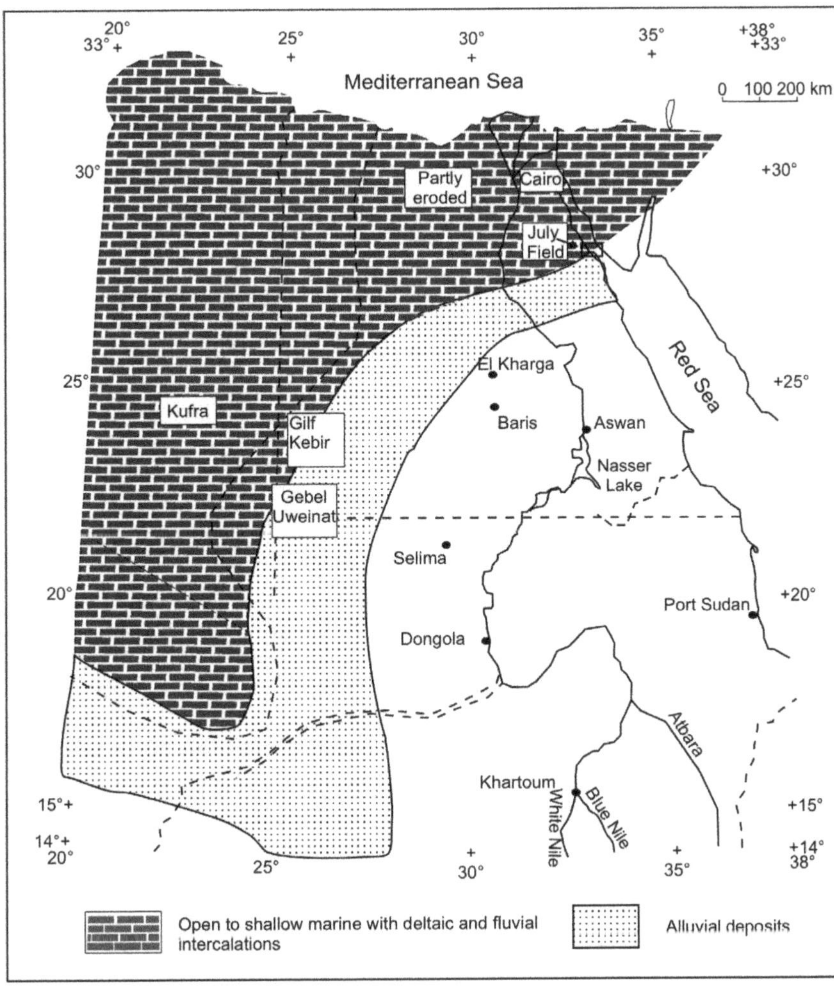

Figure 4. Distribution and facies of Lower Carboniferous sediments in northeast Africa and approximate extension of marine influence in Devonian time, in addition to the study of the July Field [30]. The arrow refers to the July Field.

2. Materials and Methods

The current samples were chosen from minor shale interlayers embedded in extensive sandstone sequences [10,11], and the location of the samples is illustrated (Figures 5 and 6). A total of fifty cuts of ditch samples were collected; no core samples could be obtained from the Gulf of Suez Petroleum Company (GUPCO). The samples are ditch-cutting samples, so the risk of caving is probable. These comprise shaly sand-sourced samples, which might be from the Nubia C and the Nubia B members. The geographic position of the J62-64 AST1 well is latitude 28.14′24.54″ N and longitude 33.16′46.32″. The geographic position of the J62-86 well is latitude 37′1.34″ N and longitude 50′19.42″.

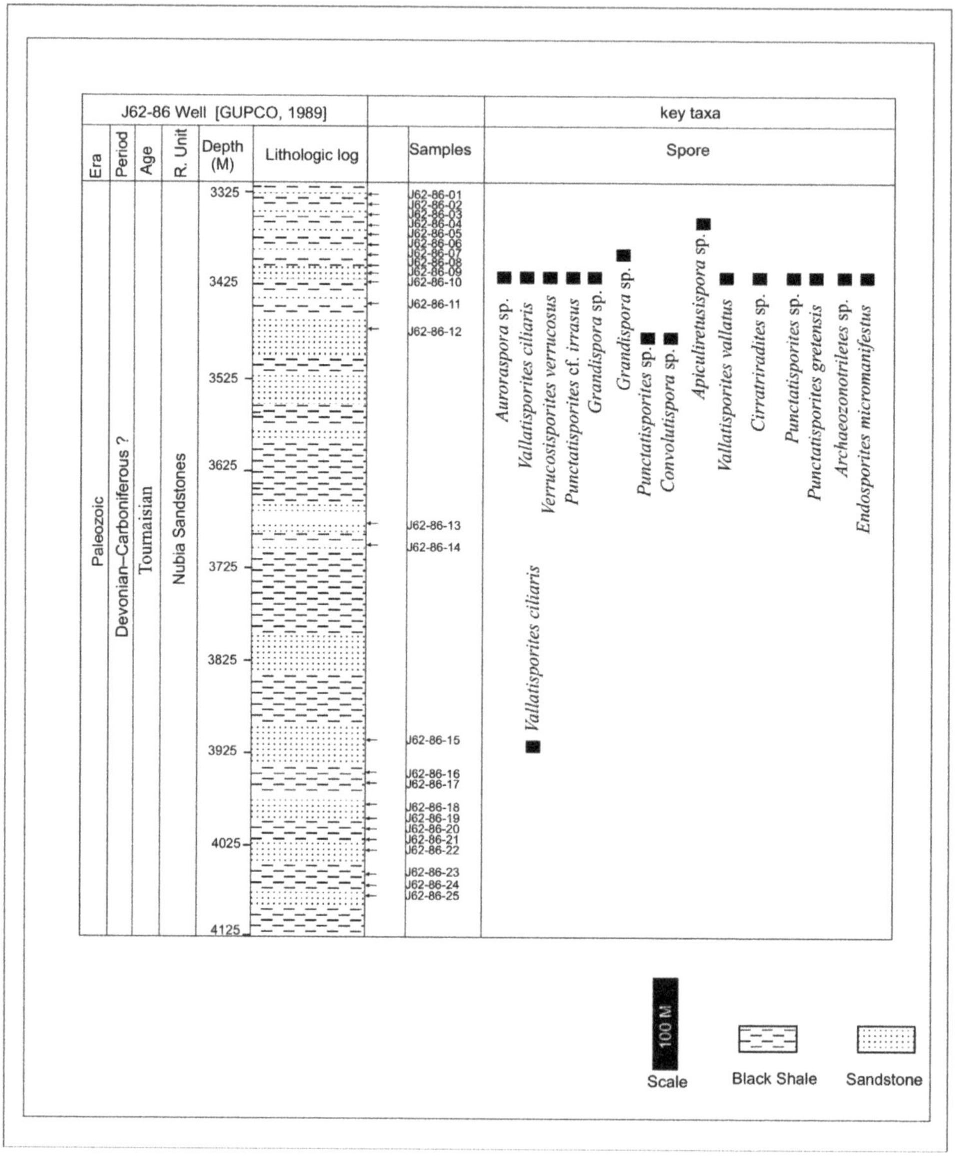

Figure 5. Lithologic log and chronostratigraphic framework of the J62-86 well, Gulf of Suez, Egypt (log based on [31]), with key taxa. The arrows refer to the location of the studied samples.

The ditch samples were initially described and studied using hand specimens and a binocular optical microscope. The lithologic analysis performed under a light microscope reveals that the obtained samples are predominantly composed of very fine sediments composed of dark black shale and sandstone. The samples were prepared for the palaeopalynological analysis according to the standard tools [33]. The obtained samples noticed that it was affected by the oil, not knowing if it was from the reservoir intervals. The samples were washed with acetone before acid treatments. The sample treatment (5 g) involved HCI-HF-HCL processing and sieving through 10 µm meshes. Some samples were also macerated in nitric acid, but the preservation has still not improved. Slide preparation included the use of ICL's Elvacite® 4345: DuPont America, Wilmington, DE, USA, and some slides used glycerin jelly.

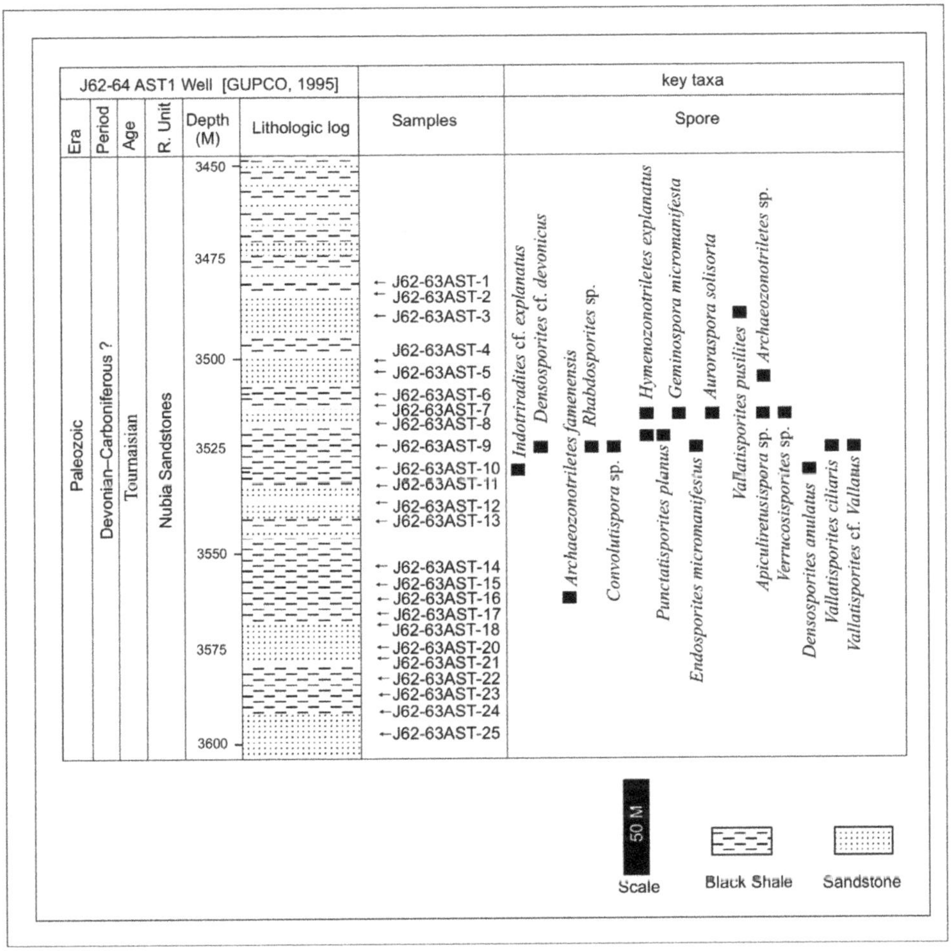

Figure 6. Lithologic log and chronostratigraphic framework of the J62-64 AST1 well, Gulf of Suez, Egypt (log based on [32]), with key taxa. The arrows refer to the location of the studied samples.

The July Field samples were prepared for conodonts. Each sample was washed with water and put on a hot plate for two hours, then left to be returned to the lab at the natural temperature. Each sample sieve was investigated under a light Optika microscope. Conodont elements were observed and isolated for SEM. Most isolated samples were missing the main taxonomic features for the identification process, which might be because of the drilling process. An isolated conodont element was tested with HCL; however, no effect was detected. Microfossils designated for SEM examination were affixed on aluminium stumps, subsequently coated with platinum via a magnetron sputter coater (JUC-5000, JEOL, Tokyo, Japan), and scrutinized under an SEM (JSMIT100, JEOL). JEOL Ltd., Akishima, Tokyo, JEOL. These SEM observations were conducted at Japan's Shizuoka Museum for Natural and Environmental History. Residual materials, slides, and original cutting samples are securely housed at the Department of Geology, Al-Azhar University (Assiut), Egypt. A list of recorded taxa is provided in Appendix A. The present manuscript is attached with the Supplementary File.

3. Results

The current investigation of the dispersed spore assemblage has studied the spore assemblage of the Nubia Sandstone deposits in the J62-86 well and the J62-64 AST1 wells (Figures 3 and 4) to assign a palynostratigraphic framework for the Middle Palaeozoic succession. The spore succession in the studied wells is based on ditch cutting with the probability of downhole contamination. The results obtained have partially been supported by conodont microfossils but without identification because the logging tools destroyed most of the taxonomic features. The palynological and conodont assemblages are illustrated and represented in Figures 5–10.

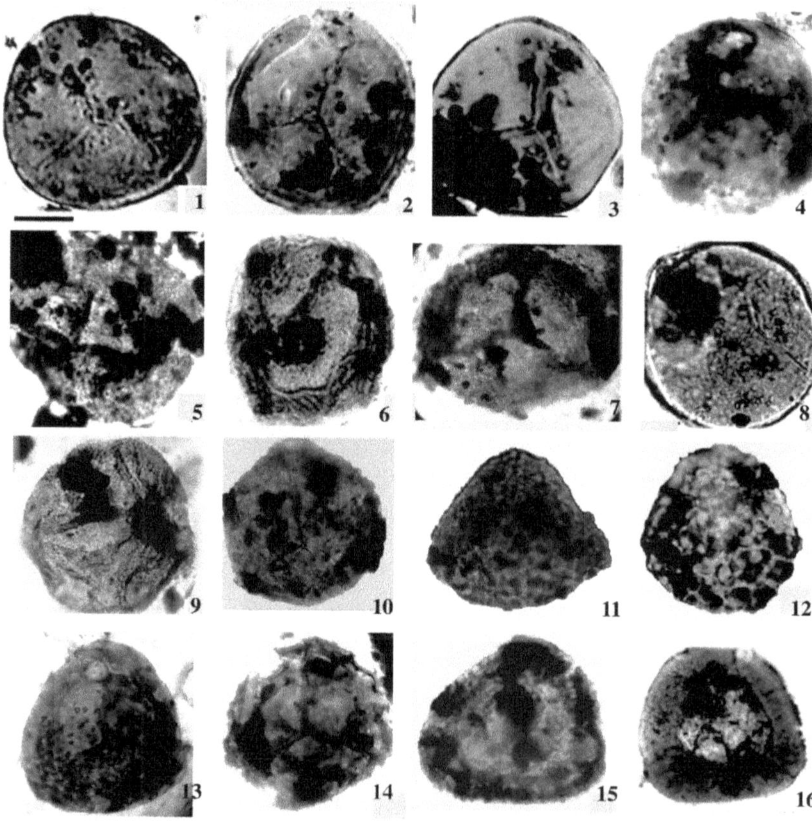

Figure 7. 1. *Punctatisporites gretensis* Balme and Henelly; J62-64 AST1 well. 2. *Punctatisporites* sp.; J62-64 AST1-012-D2: EF. V26/2. 3. *Punctatisporites planus* Hacquebard; J62-86-09-D2: EF. C42/3. 4. *Punctatisporites* sp. 5. *Verrucosisporites* sp.; J62-64 AST1-08F1. 6. *Verrucosisporites* sp.; J2-07-1. 7. *Apiculiretusispora arabiensis* Al-Ghazi. 8. *Punctatisporites* cf. *irrasus* Hacquebard; J62-86-010-1. 9. *Punctatisporites* sp.; J2-03-1. 10. *Retusotriletes* cf. *incohatus* Sullivan. 11. *Pustulatisporites* sp. 12. *Verrucosisporites verrucosus* (Ibrahim) Ibrahim; J62-86-09-AZ-16. 13. *Microreticulatisporites* sp. 14. *Retusotriletes warringtonii* Richardson and Lister. 15. *Densosporites* cf. *devonicus* Richardson; J62-86-10-1. 16. *Densosporites anulatus* (Loose) Schopf, Wilson and Bentall; J62-64 AST1-010-D1: EF. D40. Scale bar = 10 μm.

The samples in our study were obtained from the deepest wells in the Gulf of Suez, which were significantly influenced by the reservoir rock, leading to poor preservation despite multiple extraction attempts. To improve quality, we remacerated some samples and captured new photomicrographs, which were incorporated into the manuscript.

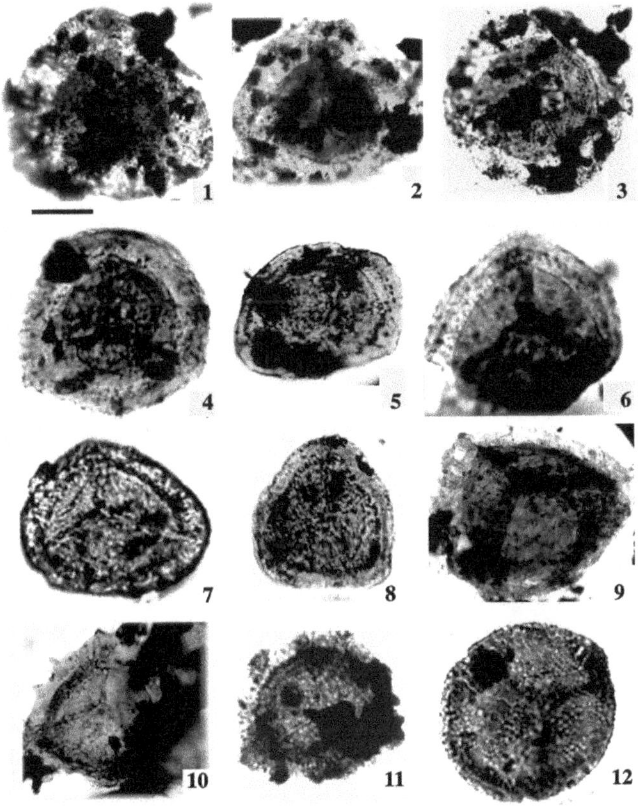

Figure 8. 1. *Auroraspora solisorta* Hoffmeister, Staplin and Malloy; J62-64 AST1-08-6 and J1-08-A1: EF. C20/2. 2. *Indotriradites* cf. *explanatus* (Luber) Playford; J62-64 AST1-010-A: EF. F51. 3. *Auroraspora* sp.; J62-86-010-1. EF. F51. 4. *Rhabdosporites* sp.: J62-86-09-B-6: P39/4 and J2-020-L1: EF. X1. 5. *Vallatisporites* sp.; J62-86-09-D1-1. 6. *Vallatisporites* sp.; J62-86-09-B-6: EF. E34. 7. *Geminospora* cf. *lemurata*; J62-86-09-E-2: EF. J51/4. 8. *Geminospora* sp.; J62-86-09-5: EF. C60-4. 9. *Vallatisporites* cf. *pusillites*. 10. *Ancyrospora* sp.; J62-64 AST1-012. 11. *Cirratriradites* sp.; J62-86-09-A-9: EF. B36. 12. *Archaeozonotriletes* cf. *famenensis* Naumova; J62-86-016-1. Scale bar = 10 μm. The samples in our study were obtained from the deepest wells in the Gulf of Suez, which were significantly influenced by the reservoir rock, leading to poor preservation despite multiple extraction attempts. To improve quality, we remacerated some samples and captured new photomicrographs, which were incorporated into the manuscript.

The studied samples from the J62-64 AST1 well and J62-86 well exhibit a diverse range of sporomorphs, with very few palynomorphs. The palynological investigation of the July Field reveals that the preservation of spore assemblages is moderate to poor (dark brown), with productive samples being concentrated in specific intervals (Figures 5 and 6). Out of fifty processed samples from the July Field in the Gulf of Suez, only a few were productive and yielded spores. The accurate stratigraphic setting of the studied samples from the July Fields will rely on the spore assemblage. The poor preservation of conodont elements prevents firm identification, but they are presumed to be of the same age as the recovered spore assemblage.

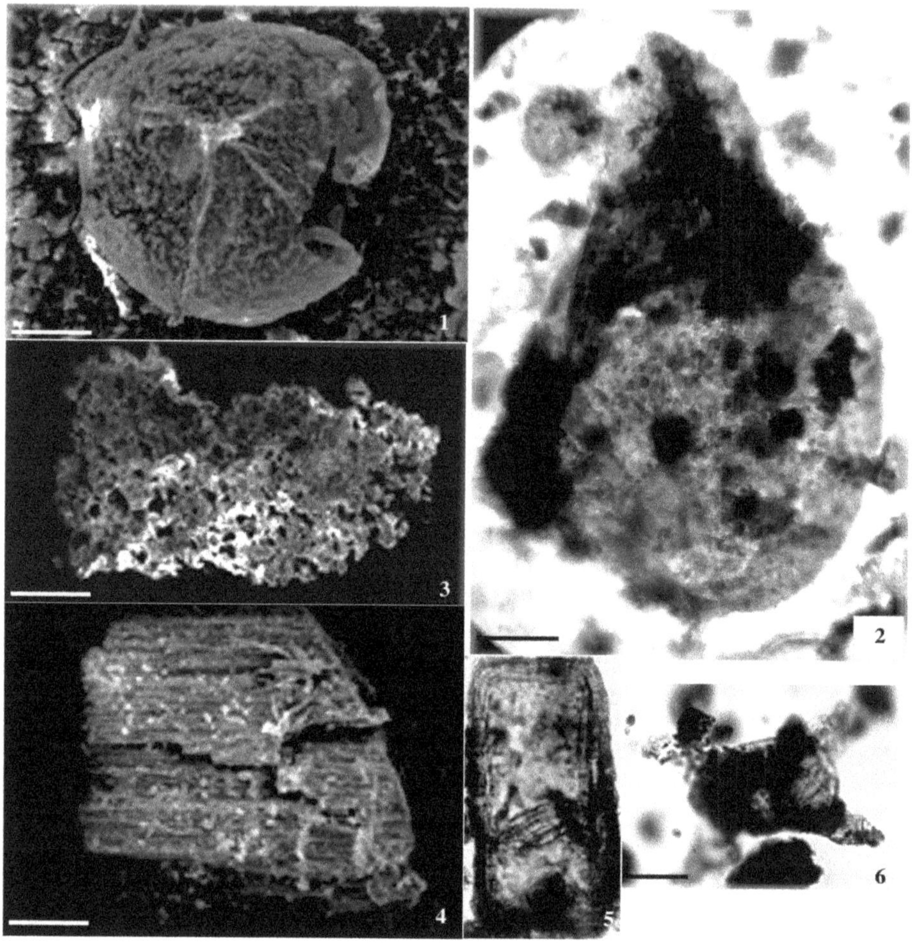

Figure 9. 1. Trilete unidentified megaspore, SEM, scale bar = 100 μm. 2. Trilete megaspore of the *Lagenoisporites* type. Note the prominent apical gula (upper structure), SEM, scale bar = 100 μm. 3. Cuticle-like sheet or pseudocellular cuticle, SEM, scale bar = 100 μm. 4. Wood fragment, SEM, scale bar = 100 μm. 5. Filamentous green algae; J2-05-1: EF. G9, scale bar = 10 μm. 6. *Schizocystia bicornuta* Jardinè; J62-86 well, scale bar = 10 μm. The samples in our study were obtained from the deepest wells in the Gulf of Suez, which were significantly influenced by the reservoir rock, leading to poor preservation despite multiple extraction attempts. To improve quality, we remacerated some samples and captured new photomicrographs, which were incorporated into the manuscript.

Dispersed spore assemblages occur only in some restricted intervals and not in the whole profile (Figures 3 and 4). The most productive samples in J62-64 AST1 well were at depths [3500, 3525, and 3530 m] of samples [8,10,12]. The productive interval in J62-86 well is [3425, and 3450] of samples [9,10]. Zonate and camerate spores, microgranulate and verrucate spores of the genus *Verrucosisporites* (Ibrahim) Smith and Butterworth, and spores of the genus *Hymenozonotriletes* Naumova dominated. Moreover, some megaspores were observed in the genus. Abundant are plant fragments, tracheids, pseudocellular cuticles, cuticles of arthropods, and tubular fragments. Sculptured spores of the genus *Apiculiretusispora* (Streel) Streel laevigate and patinate *Archaeozonotriletes* (Naumova) Allen, crassitate forms including *Aneurospora* (Streel) Richardson et al., and ornamented patinate spores (e.g., *Cymbosporites* Allen) are rare. The spore assemblage is characterized mainly by spores of genera *Apiculatisporites* (Ibrahim) Potonié and Kremp, *Convolutispora* Hoffmeister et al., *Densosporites* (Berry) Butterworth et al., *Geminospora* (Balme) Playford, *Grandispora* Hoffmeister, Staplin and Malloy, *Punctatisporites* Ibrahim, *Pustulatisporites* Potonié and Kremp *Rhabdosporites* Richardson, *Samarisporites* Richardson, *Spelaeotriletes* Neves and Owens, *Vallatisporites* (Hacquebard) Sullivan, *Chelinospora* Allen, and *Synorisporites* Richardson and Lister, some *Reticulatisporites* Ibrahim, and several representatives of *Leiosphaerida* Eisenack, *Lophosphaeridium* Timofeev, *Schizocystia bicornuta* Jardiné, and fragments of filamentous green algae.

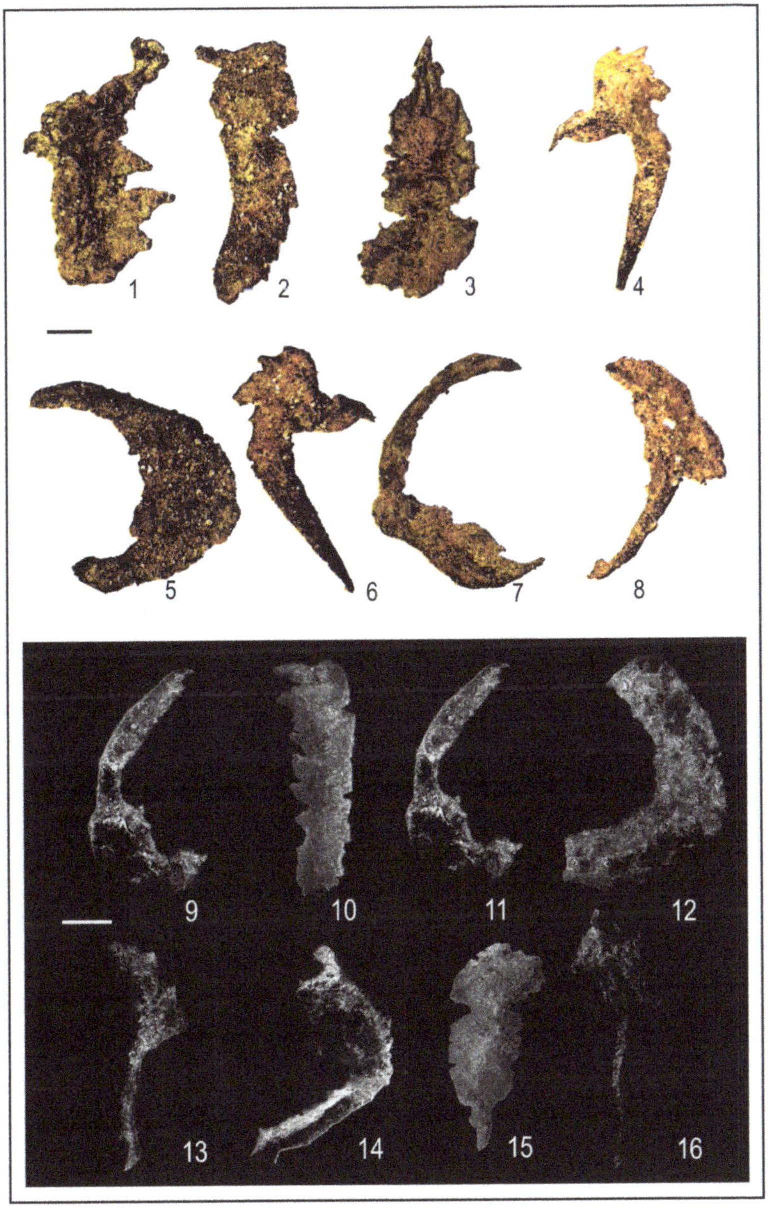

Figure 10. The drilling process caused unidentified conodont assemblage from 1 to 7. Photomicrographs of conodont fragments: 8–16. Conodont fragments: 1–7 (Scale bar 400 μm).

The palynofacies evidence suggests that the spore assemblage was deposited near marine settings, as indicated by conodonts. Additionally, the dominance of plant remains provides further evidence of the near source of the vegetational cover.

4. Discussion

4.1. The Stratigraphical Position of Spore Assemblage in the Gulf of Suez, Egypt

The July Field spore assemblage contains marker spore taxa of the Devonian and the Carboniferous ages. The unconformity surface between the Nubia "C" and Nubia "B" in (Figure 1) is interesting because it could bear new information for the missing ages of the Devonian or the Devonian–Carboniferous boundary.

The Palaeozoic marine transgressions that occurred during the Early to Middle Palaeozoic accelerated the reworking and extensive dispersal of siliciclastic sediments [3]. The distribution of the Lower Carboniferous sediments in Northeast Africa and the extension of marine influence in the Devonian time is illustrated in (Figure 4), which declares that the study of the July Field area was in an open, too shallow marine environment with deltaic and fluvial interactions [30].

The July Field spore assemblage suggests being younger than the Uppermost Devonian age. This is because [34,35] concur that stratigraphically important *Retispora lepidophyta* (Kedo) Playford is classified to the Uppermost Devonian, and it is not documented in the July Field spore assemblage. However, the appearance of *Verrucosisporites nitidus* (Naumova) Playford defines the base of the Uppermost Devonian *Retispora lepidophyta–Verrucosisporites nitidus* (LN) Biozone in Western Europe, with *Vallatisporites verrucosus* (Ibrahim) Ibrahim typically evolving at the same stratigraphic level [35]. *Verrucosisporites nitidus* has been regarded as the most recent Devonian characteristic microspore associated with *Retispora lepidophyta* [36]. Spore assemblage exhibits similarities to that noted from the latest Devonian and Early Carboniferous microspore assemblage in Saudi Arabia [35]. The assemblage zone characterized by *Verrucosisporites nitidus* and *Vallatisporites verrucosus*, as documented by [37], is situated within the (Late Strunian) Late Famennian age. This specific zone is distinguished by the prevalence of characteristic species that exhibit occurrences in combination with the taxa currently recorded from the July Field. The species of note within this assemblage include *Verrucosisporites nitidus*, *Vallatisporites verrucosus*, and *Indotriradites* cf. *explanatus* (Luber) Playford.

The microspores obtained from the July Field can be correlated with the *Vallatisporites pusillites–Retusotriletes incohatus* assemblage zone [37]. This zone has been dated to encompass the Tournaisian age. The current spore assemblage from the July Field shares the occurrence of genera, including *Auroraspora* Hoffmeister et al.; *Archaeozonotriletesi* cf. *famenensis* Naumova, *Grandispora*, *Archaeozonotriletes*; *Retusotriletes* cf. *incohatus*; and *Hymenozonotriletes*. however, the associated taxa between the [37] and the July Field wells in the current study recommend the Tournaisian age.

The July Field assemblage is characterized by the appearance of the monopseudosaccate spore without foveoreticulate exoexine, such as *Auroraspora*, and *Grandispora*, which is considered the description of the *Rugospora flexuosa–Grandispora cornuta* assemblage zone [37]; this represents partial correlation.

The Givetian–Frasnian boundary deposits in Byelarus were proposed to be distinguished by a marked palynofloral shift [38,39]. *Verrucosisporites nitidus* is recognized as a marker species indicative of the Devonian–Carboniferous boundary [40,41]. The record of this species in the current spore assemblage in the July Field suggests that the July Field could be located between the Nubia "C" and the Nubia "B. This refers to the probability of the Devonian–Carboniferous transition interval that occurred in the July Field.

Ref. [42] investigated the Carboniferous assemblages from the Kingdom of Saudi Arabia and identified them as Tournaisian, mainly characterized by *Indotriradites* cf. *explanatus*, *Auroraspora macra* Sullivan, *Punctatisporites irrasus* Hacquebard, and various other taxa, including *Calamospora* Schopf et al., *Knoxisporites literatus* (Waltz) Playford, *Perotrilites perinatus* Hughes and Playford, *Retusotriletes* (Naumova) Streel, *Vallatisporites vallatus* Hacquebard, and *V. verrucosus*. These spores are also recorded in the studied samples from the July Field, with the absence of the uppermost Devonian taxon *Retispora lepidophyta* being unobserved both in the July Field and the assemblage documented by [38]. Furthermore, the Early Carboniferous spore and pollen taxa noted from AR'AR-1 well by [42] were not represented in the July Field study.

The Gondwana Devonian and Carboniferous palynological records reveal crucial insights into its paleobiogeographical and palaeoclimatic evolution [43–45].

The assemblage zone A, recorded from the Carboniferous strata of the Abu Rodeiyim boreholes [24] in West-Central Sinai, Egypt, assumes significant resemblance to the spore assemblage currently documented from the samples analysed in the July Field. There are shared occurrences of *Punctatisporites*, *Verrucosisporites*, *Convolutispora*, *Vallatisporites*, and *Densosporites*, and this assemblage has been assigned to the Late Viséan age, primarily in regions encompassing North Africa, Western Europe, and Western Australia.

The observation of the megaspore in the July Field miospore assemblage is considered to represent an indication that it might indicate the presence of arborescent bisporangiate lycophytes [46,47] in the Gulf of Suez, Egypt. Trilete gulate megaspores of the *Lagenoisporites* type are recorded among other megaspores.

The record of *Schizocystia bicornuta* in the July Field wells was recorded from the Upper Devonian of Algeria as a guide fossil [48]. Filamentous green algae observed in the July Field studied samples have great similarity to the individual specimens morphotypes reported from the Early Devonian Rhynie chert, which inhabited freshwater ponds/lakes [49]. This assumption can greatly agree with the new reconstructed palaeodepositional environment for the July Field inferred from [30], which indicates that the area was of the deltaic, fluvial, and alluvial origin. The presence of cuticle-like sheets, cuticles of arthropods, plant remains, and tracheids are consistent with this assumption. The occurrences of genera *Lophosphaeridium* and *Leiosphaerida* are indicated in the shallow water environment. The July Field miospore assemblage could be assumed to have been transported from the vegetation source by the deltaic system.

Comparison with Other Dispersed Spore Assemblages

The recorded spore assemblage from the currently studied July well samples is correlated with the areas from the Gondwana and the Old Red Sandstone continent and adjacent regions [37]. The next section will discuss the correlation, and the established zonation is constructed in (Figure 11). The current spore assemblage is correlated with the Old Red Sandstone continent [37].

The documented spore assemblages from the Western Argentine and Southern Bolivian regions, dating to the Middle Devonian age, are indicative of the established patterns within the Afro-South American Subrealm [37]. The identification of *Grandispora pseudoreticulata* in the present investigation of the July Field from the Gulf of Suez, Egypt, serves to corroborate the hypothesis that the extended patterns of the Afro-South American Subrealm encompass Egypt within these palaeogeographical contexts of Gondwana [43]. This taxon, *Grandispora pseudoreticulata*, has also been reported from the Middle Devonian strata of Northwestern Argentina [38]. Furthermore, it has been documented within the Mid-Late Devonian assemblages of herbaceous lycophytes from Northern Argentina and Bolivia [43]; however, the infrequent presence of some taxa does not help to establish a correlation [44]. This indicates that the biogeographical patterns of the Afro-South American Subrealm could extend to include or reach Egypt within these palaeogeographical regions of Gondwana [44].

The comparative analysis with the palynostratigraphic studies conducted by [37] is related to the present spore assemblage acquired from the July Field in the Gulf of Suez, Egypt. The *Vallatisporites pusillites sensu lato* and *Retispora lepidophyta* zone, as characterized by [37], have been attributed to a Late Famennian (Strunian) age. This assemblage is characterized by the presence of *Vallatisporites pusillites*, and dominance of *Vallatisporites* demonstrates notable similarities to the findings presented in the current study of the July Field. The *Vallatisporites pusillites* encountered in the present study are well documented across numerous locations, including various regions in North America, such as Western New York State, Pennsylvania, and international locations, including Belgium, Kazakhstan, Tibet, South China, the former U.S.S.R., Eastern Europe, and Western Europe [37].

Indotriradites cf. *explanatus*, previously recorded from locations in Ireland, Southern Ireland, the Ardennes–Rhine region, Western New York State, and Northern Pennsylvania [37], has also been identified in the current study conducted at the July Field.

In addition to another zone delineated by [37], the *Verrucosisporites nitidus–Vallatisporites verrucosus* assemblage zone corresponds to the Famennian (Strunian) age. This assemblage is distinguished by the presence of *Verrucosisporites nitidus, Vallatisporites verrucosus*. A comparable occurrence of *Verrucosisporites riitidus, Vallatisporites verrucosus, Discernisporites, Grandispora, Indotriradites* cf. *explanatus*, and *Knoxisporites* has been documented in the present study conducted in the July Field.

The abundance of the *Vallatisporites* in the present study can highly recommend the transition between the Devonian and Carboniferous periods. However, several European localities noted the *lepidophyta* group's extinction [45]. The non-recorded *lepidophyta* group in the July Field could be caused by some changes in the palaeoenvironment of the deposits. The disappearances of the *Retispora lepidophyta* were proposed to be of the Devonian–Carboniferous characteristic boundary [50]. We could not confirm that similar conditions controlled the Hangenberg Shales [45] if extended to the July Field. However, the infrequent occurrences of marine micophytoplankton in the July Field could be questionable.

Apiculiretusispora plicata (Allen) Streel was recorded from the palynological assemblage of the Lower Devonian of China [51,52]. *Geminospora libyensis* Richardson was recorded from the Devonian spores of South China, Saudi Arabia, and Australia [53]. The occurrence of megaspores of the *Lagenoisporites* type is important for the spore evolution in the Devonian–Carboniferous of Egypt and means the occurrence of first arborescent lycophytes, i.e., bisporangiate members of the *Lepidophloios* family [54] in Egypt. The megaspore was proposed to be regarded as the event that occurred to coincide with the first occurrence of the distinctive group of Famennian species of *Diducites* [55]. The affinity of this important megaspore is attributed to Lepidodendraceae [56]. *Apiculiretusispora arabiensis* Al-Ghazi was recorded from the Emsian of Saudi Arabia [57] and from Lower Devonian Rañeces—La Vid groups of Northern Spain [58].

The location of the study of the July Field (Figures 3 and 4) in the distribution facies of the Lower Carboniferous sediments in Northeast Africa and the extension of the marine influence in the Devonian age [30] illustrating the palaeodepositional setting of the study area of the July Field are pictured. The July Field palynological assemblage contains spores, rare marine microphytoplankton, plant fragments, and cuticle sheets. In addition to the presence of the conodont elements. This July Field assemblage agrees with the facies of the deltaic and fluvial intercalations of the Devonian age extended in North Africa, including the Gulf of Suez, Egypt. The spore assemblage is greatly represented by the dominance of the early lycophytes and some bryophytes or early tracheophytes, with the suggestion of the emergence of the early ferns.

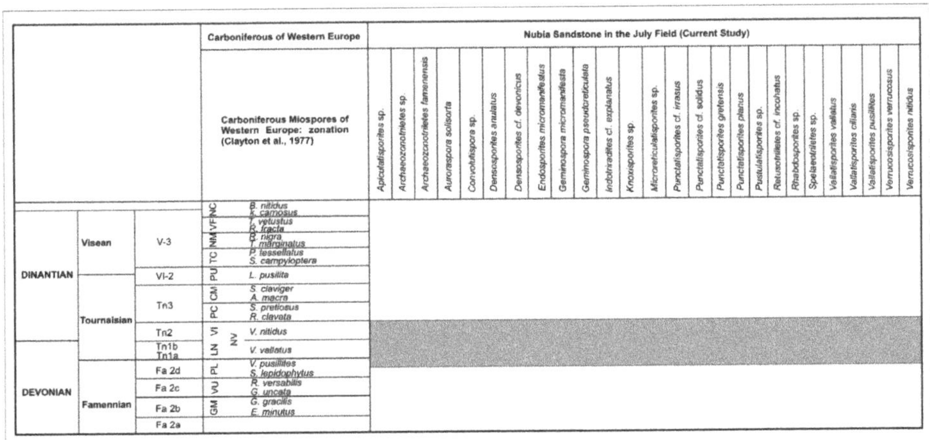

Figure 11. Correlation of the recorded spore assemblage with the zone of the Old Red Sandstone continent and adjacent regions, modified from [59]. The grey box refers to the correlated zone of the Carboniferous of Western Europe with the July Field in the Gulf of Suez (Current study), Egypt.

4.2. The Tournaisian Spore Assemblage in the Gulf of Suez, Egypt

The correlation of the present spore assemblage obtained from the July Field in the Gulf of Suez can be correlated to the *Vallatisporites vallatus–Retusotriletes incohatus* (VI) Subzone [59]. The present July spore assemblage is correlated with the palynological Western Europe zonation [59]. The *Vallatisporites pusillites–Spelaeotriletes lepidophytus* (PL) Zone is characterized by the presence of *Spelaeotriletes lepidophytus* (Kedo) Streel and *Vallatisporites pusillites* (Kedo) Dolby and Neves, while *Spelaeotriletes lepidophytus* is very abundant [59]. These spore taxa are recorded from the present study in the July Field wells. Another significant feature that the spore assemblage recorded in the (PL) zone of [59] is containing poorly preserved spores of the Devonian, while the well-preserved spores are thought to belong to the Carboniferous in Tournaisian age [59], this feature is also noticed in the July Field. The Devonian–Carboniferous boundary was thought to be in the (PL) zone, sharing the same idea for the present July spore assemblage.

There is a correlation with the *Verrucosisporites nitidus–Vallatisporites vallatus* (NV) zone [59] with the present July spore assemblage containing *Indotriradites* cf. *explanatus* with *Spelaeotriletes lepidophytus* (Kedo) Streel, *Vallatisporites pusillites*, and *Rugospora flexuosa*. The (NV) zone [59] shares the same taxa with the July spore assemblage.

There is a correlation with the *Spelaeotriletes lepidophytus–Verrucosisporites nitidus* (LN) subzone containing *Spelaeotriletes lepidophytus*, *Retusotriletes incohatus* Sullivan, *Verrucosisporites nitidus*, *Raistrickia macrura* (Luber) Dolby and Neves, *Raistrickia variabilis* Dolby and Neves, *Dictyotriletes submarginatus* Playford, *Hymenozonotriletes commutatus* Naumova, *Lophozonotriletes malevkensis* Kedo, *Vallatisporites pusillites*, *Vallatisporites verrucosus*, *Auroraspora macra* Sullivan, *Auroraspora poljessica* (Kedo) Streel, *Discernisporites micromanifestus* (Hacquebard) Sabry and Neves, and *Rugospora flexuosa*. The (LN) subzone of [59] is sharing the findings of the main spore taxa in the (LN) subzone except for the non-record of *Raistrickia macra*, *Raistrickia variabilis*, *Dictyotriletes submarginatus*, *Hymenozonotriletes commutatus*, *Lophozonotriletes malevkensis*, *Vallatisporites pusillites*, *Vallatisporites verrucosus*, *Auroraspora macra*, *Auroraspora poljessica*, *Disernisporites micromanifestus*, and *Rugospora flexuosa*.

The base of the VI subzone is defined from the Munster Basin, Germany, the Kinsale Formation, and it represents the upper part of the "Old Red Sandstone" and the lower part of the Porter's Gate Formation at Hook Head, Co. Wexford, and the lower part of the Lower Limestone Shales in the Mendips, S.W. Britain, UK [59]. This subzone is defined by the disappearances of *Spelaeotriletes lepidophytus* Owens and Streel together with *Vallatisporites pusillites* Hacquebard and *Rugospora flexuosa* Higgs [59]. The lowermost assemblage of this subzone is dominated by smooth acamerate spores *Punctatisporites irrasus* Hacquebard and *Retusotriletes incohatus* Hacquebard, *Lophozonotriletes cristifer* (Luber) Kedo, *Retusotriletes triangulatus* (Streel) Streel, *Pulvinispora quasilabrata* Higgs, *Raistrickia corynoges* Sullivan (Naumova) Playford, *Raistrickia corynoges* Sullivan, *Convolutispora* sp., *Corbulispora cancellata*, *Dictyotrilletes submarginatus*, *Indotriradites* cf. *explanatus*, *Knoxisporites literatus* (Waltz) Playford, *Lophozonotriletes cristifer* (Luber) Kedo, *Lophozonotriletes malevkensis* Birina, *Lophozonotriletes rarituberculatus* (Luber) Kedo, and *Latosporites* sp. *Dibolisporites distinctus* (Clayton) Playford appears but is not common. *Spelaeotriletes balteatus* (Playford) Higgs is frequently abundant [59]. The listed taxa of the (VI) subzone of [59] are recorded in the present spore assemblage in the July Field in the Gulf of Suez, Egypt, except some taxa are not observed. These are *Spelaeotriletes lepidophytus*, *Rugospora flexuosa*, *Lophozonotriletes cristifer*, *Pulvinispora quasilabrata*, *Raistrickia corynoges*, *Corbulispora cancellata*, *Dictyotrilletes submarginatus*, *Lophozonotriletes cristifer*, *L.s malevkensis*, *L. rarituberculatus*, *Dibolisporites distinctus*, and *Spelaeotriletes balteatus*.

The correlation with the Carboniferous spore zonation [59] shares many similarities to the present July spore assemblage: these zones are the *Vallatisporites pusillites–Spelaeotriletes lepidophytus* (PL) zone, *Verrucosisporites nitidus–Vallatisporites vallatus* (NV) zone, *Spelaeotriletes lepidophytus–Verrucosisporites nitidus* (LN) subzone, and VI subzone of the *Vallatisporites vallatus–Retusotriletes incohatus*. The July study well spore assemblage is well established to be correlated to the PL zone, NV zone, LN subzone, VI subzone [59]. This confirms that the July study wells are of the Tournaisian age.

The present July spore assemblage is older than the study of the regions of Sinai and the Gulf of Suez [17,19–24], which were reported to be of the Visean to Late Carboniferous age. The presence of megaspores in the July Field spore assemblage indicates a significant evolutionary event in the palaeovegetation of Egypt. It could refer to the evolution of the seed plants, of which it could mark the diversity of plant proliferation of various groups. Reconstruction of the palaeovegetation of the Lowermost Carboniferous Nubia sandstone in the Gulf of Suez, Egypt, is introduced (Figure 12).

Figure 12. A simple reconstruction of the palaeovegetation and palaeoenvironment of the Nubia sandstone from the study of the July Field in the Gulf of Suez, Egypt.

The correlation within the upper Devonian and Carboniferous miospore assemblage suggests that the July Field spore assemblage is more suitable to be compatible with the Lowermost Carboniferous age. The stratigraphic setting of the Nubia sandstone is complex and not resolved before. The location of the Nubia "B" in the lithostratigraphic column in the Gulf of Suez (Figure 1) illustrating an unconformity surface for the lower interval of the Nubia "B" [4–11]. The observation of the Devonian and Carboniferous spore taxa in the lower interval of the study wells from the July Field indicates the Devonian age, but the July Field is palynologically more consistent to the Lowermost Carboniferous than the uppermost Devonian ages. The July Field study wells are palynologically dated to the Tournaisian age.

5. Conclusions

For the first time, the palynological investigations have enabled the description of lithostratigraphic and palynostratigraphic units within the subsurface layers of the Nubia Sandstone. The established palynostratigraphic units have been systematically correlated and temporally assigned to the Lowermost Carboniferous (Tournaisian) and could reach the transition interval between the Devonian and Carboniferous boundary. The age dating of the lower interval of the Nubia Sandstone B member, inferred from the correlation of spore assemblages, is assumed to correspond to the Devonian–Carboniferous intervals. Correlation with the (VI) subzone from S.W. Britain can mark the introductory identification of such a stratigraphic classification within the Gulf of Suez, Egypt. The chronological assessment of the analysed samples from the Nubia Sandstone is characterised as the basal interval of the Nubia B member. The stratigraphical position of spore assemblage in the Gulf of Suez, Egypt, is suggested to be located between the lower interval of the Nubia B and the upper interval of the Nubia C.

The analysed specimens have been predominantly characterized by conodont elements, yet taxonomic identification was hindered due to the presumed destructive impact of drilling tools on the morphological features of the collected specimens. The palynological and palaeontological data obtained from the July Field indicate the presence of the deposition of the Devonian–Carboniferous intervals in the Gulf of Suez, Egypt. It is suggested to be of the deltaic, fluvial, or alluvial origin. This conclusion is very important for the understanding of the paleogeography and palaeodeposition of Egypt during the Devonian age.

Supplementary Materials: The following supporting information can be downloaded at: https://www.mdpi.com/article/10.3390/life15060872/s1.

Author Contributions: Conceptualization, A.M. and J.B.; methodology, A.M. and J.B.; software, A.M.; validation, A.M. and J.B.; formal analysis, A.M. and J.B.; investigation, A.M. and J.B.; resources, A.M.; data curation, A.M.; writing—original draft preparation, A.M. and J.B.; writing—review and editing, A.M. and J.B.; visualization, A.M.; supervision, A.M. and J.B.; project administration, A.M.; funding acquisition, J.B. All authors have read and agreed to the published version of the manuscript.

Funding: This research was funded by the project of the Czech Grant Agency (GAČR) No. 21–10799S and the research plan of the Institute of Geology of the Czech Academy of Sciences, v.v.i., RVO67985831.

Institutional Review Board Statement: Not applicable.

Informed Consent Statement: Not applicable.

Data Availability Statement: Palynological data are available in the database stored at the Department of Geology, Faculty of Science, Al-Azhar University (Assiut), Egypt.

Acknowledgments: The authors wish to convey sincere appreciation to GUPCO for the provision of the samples and logs utilized in this investigation, as well as to the Egyptian Oil Ministry for granting permission for the dissemination of this work. The first author acknowledges Akihisa Kitamura and Julien Legrand from Shizuoka University, Japan, for their support and scientific advice. The Museum of Natural and Environmental History of Shizuoka, Japan, is duly acknowledged for allowing the scanning electron microscopy (SEM) observations conducted in this study. We thank Annalisa Ferretti (Department of Chemical and Geological Sciences, University of Modena and

Reggio Emilia, Via Campi 103.), Petra Tonarová (Czech Geological Survey, Prague, Czech Republic), Jerzy Dzik (Institute of Paleobiology, Polish Academy of Sciences, Warsaw, Poland), and Mahmoud Kora (Geology, Mansoura University, Egypt) for their help and opinions in the identification of the conodont material.

Conflicts of Interest: The authors declare they have no competing interests.

Appendix A

List of recorded taxa from the July studied (J62-86 and J62-64 AST1) wells in the Gulf of Suez, Egypt.

Spore

Ancyrospora sp. Figure 8-10
Aneurospora sp.
Apiculatisporites sp.
Apiculiretusispora arabiensis Al-Ghazi. Figure 7-7
Apiculiretusispora sp.
Archaeozonotriletes cf. *famenensis* Naumova. Figure 8-12
Archaeozonotriletes sp.
Auroraspora solisorta Hoffmeister Staplin and Malloy. Figure 8-1
Auroraspora sp.
Chelinospora sp.
Cirratriradites sp. Figure 8-11
Convolutispora sp.
Cristatisporites sp.
Cymbohilates sp.
Cymbosporites sp.
Densosporites anulatus (Loose) Smith and Butterworth. Figure 7-16
Densosporites cf. *devonicus* Richardson. Figure 7-15
Densosporites sp.
Diaphanospora sp.
Endosporites sp.
Geminospora cf. *lemurata* Figure 8-7
Geminospora micromanifesta (Naumova) Owens.
Geminospora libyensis Richardson
Geminospora sp.
Grandispora pseudoreticulata (Menéndez et al.) Ottone
Grandispora sp.
Indotriradites cf. *explanatus* (Luber) Playford. Figure 8-2
Hymenozonotriletes sp.
Indotriradites sp.
Knoxisporites sp.
Leiotriletes sp.
Microreticulatisporites sp. Figure 7-13
Punctatisporites cf. *irrasus* Hacquebard. Figure 7-8
Punctatisporites cf. *solidus* Hacquebard
Punctatisporites gretensis Balme and Henelly. Figure 7-1
Punctatisporites planus Hacquebard. Figure 7-3
Punctatisporites sp. Figure 7-2
Pustulatisporites sp. Figure 7-11
Retusotriletes sp.
Retusotriletes cf. *incohatus* Sullivan. Figure 7-10

Retusotriletes warringtonii Richardson and Lister. Figure 7-14
Rhabdosporites sp. Figure 8-4
Samarisporites sp.
Samarisporites triangulatus Allen
Spelaeotriletes sp.
Synorisporites sp.
Synorisporites verrucatus Richardson and Lister
Vallatisporites cf. *vallatus* Hacquebard.
Vallatisporites ciliaris (Luber) Sullivan
Vallatisporites pusillites (Kedo) Dolby and Neves.
Vallatisporites sp. Figure 8-5
Vallatisporites vallatus Hacquebard.
Vallatisporites verrucosus Hacquebard.
Verrucosisporites nitidus (Naumova) Playford
Verrucosisporites sp. Figure 7-5
Verrucosisporites verrucosus (Ibrahim) Ibrahim. Figure 7-12
Megaspore
Lagenoisporites type. Figure 9-2
Microphytoplankton
Leiosphaerida sp.
Lophosphaeridium sp.
Schizocystia bicornuta Jardinè. Figure 9-6.
Fungi
Filamentous green algae. Figure 9-5
Non-spore
Cuticle-like sheet. Figure 9-3.
Cuticle of arthropods
Plant remains. Figure 9-4.
Tracheid

References

1. Weissbrod, T. A reassessment of the Naqus Formation in Sinai and the Eastern Desert of Egypt: Stratigraphic and tectonic implications. *Isr. J. Earth Sci.* **2004**, *53*, 87–97. [CrossRef]
2. Issawi, B.; El-Hinnawi, N.M.; El Khawaga, L.; Labib, S.; Anani, N. Contribution to the geology of Wadi Feiran area, Sinai, Egypt. *Egypt. Geol. Surv. Rep.* **1981**, *3*, 43.
3. El Hawat, A.S. Sedimentary basins of Egypt: An overview of dynamic stratigraphy. In *African Basin. Sedimentary Basin of the World*; Selley, R.C., Ed.; Elsevier: Amsterdam, The Netherlands, 1997; Volume 3, pp. 39–85.
4. Hempton, M.R. Constraints on Arabian Plate motion and extensional history of the Red Sea. *Tectonics* **1987**, *6*, 687–705. [CrossRef]
5. Joffe, S.; Garfunkel, Z. Plate kinematics of the circum Red Sea-a re-evaluation. *Tectonophysics* **1987**, *141*, 5–22. [CrossRef]
6. Salah, M.G. Geochemical evaluation of the southern Gulf of Suez, Egypt. In Proceedings of the 11th Petroleum Exploration and Production Conference, Cairo, Egypt, 7–10 November 1992; pp. 383–395.
7. Darwish, M.; El-Araby, A. Petrography and diagenetic aspects of some siliciclastic hydrocarbon reservoirs about the rifting of the Gulf of Suez, Egypt. *Geol. Soc. Spec. Publ.* **1993**, *1*, 155–187.
8. Patton, T.L.; Moustafa, A.R.; Nelson, R.A.; Abdine, S.A. Tectonic Evolution and Structural Setting of the Suez Rift: Chapter 1: Part I. In *Interior Rift Basins*; Landon, S.M., Ed.; AAPG Special Publications Memoir: Tulsa, OK, USA, 1994; Volume 59, pp. 9–55.
9. Alsharhan, A.S. Petroleum geology and potential hydrocarbon plays in the Gulf of Suez Rift Basin, Egypt. *AAPG Bull.* **2003**, *87*, 143–180.
10. Schütz, K.I. Structure and Stratigraphy of the Gulf of Suez, Egypt: Chapter 2: Part I. Type Basin: Gulf of Suez. In *Interior Rift Basins*; Landon, S.M., Coury, A.B., Eds.; AAPG Memoir: Tulsa, OK, USA, 1994; pp. 57–96.
11. Alsharhan, A.S.; Salah, M.G. Lithostratigraphy, sedimentology and hydrocarbon habitat of the pre-Cenomanian Nubian Sandstone in the Gulf of Suez oil Province, Egypt. *GeoArabia* **1997**, *2*, 385–400. [CrossRef]
12. Soliman, S.M.; El Fetouh, M.A. Carboniferous of Egypt: Isopach and lithofacies maps. *AAPG Bull.* **1970**, *54*, 1918–1930. [CrossRef]

13. Kostandi, A.B. Facies maps for the study of the Paleozoic and Mesozoic sedimentary basins of the Egyptian region. In Proceedings of the 1st Arab Petroleum Congress, Cairo, Egypt, 16–23 April 1959; Volume 2, pp. 54–62.
14. Abdallah, A.M.; El Adindani, A.; Fahmi, N. Stratigraphy of lower Mesozoic rocks on the western side of the Gulf of Suez, Egypt. *Geol. Surv. Egypt.* **1963**, *27*, 23.
15. Weissbrod, T. The Paleozoic of Israel and Adjacent Countries (Lithostratigraphic Study). Ph.D. Thesis, The Hebrew University of Jerusalem, Jerusalem, Israel, 1980; 275p. (In Hebrew, English Abstract)
16. Issawi, B.; Jux, U. Contributions to the stratigraphy of the Paleozoic rocks in Egypt. *Geol. Surv. Egypt.* **1982**, *64*, 28.
17. Synelnikov, A.S.; Kollerov, D.K. Palinologic Analysis and Age of Coal Samples from El-Bedaà-Thora District, West-Central Sinai. *EGS* **1959**, *4*, 1–2.
18. Saad, S.I. Pollen and spores were recently discovered in the coals of the Sinai region. 2. Um, Bogma district. *Palaeontogr. B* **1965**, *115*, 139–149.
19. Omara, S.; Schultz, G. A lower Carboniferous microflora from southwestern Sinai, Egypt. *Palaeontogr. B* **1965**, *117*, 47–58.
20. Sultan, I.Z. Carboniferous microspores from a black shales unit in the Gulf of Suez, Egypt. *Sci. Géol. Bull. Mém.* **1977**, *30*, 189–201.
21. Eames, L.E. Palynologic definition of Paleozoic unconformity-bounded sequences, Gulf of Suez region, Egypt. In *Proceedings of the EGPC 7th Exploration Seminar, EGPC, Cairo, Egypt, March 1984*; Egyptian General Petroleum Corporation: Cairo, Egypt, 1984; pp. 117–125.
22. Sultan, I.Z. Palynostratigraphy of Carboniferous rock sequence in Wadi Araba well n°1, Gulf of Suez region, Egypt. *Sci. Géol. Bull. Mém.* **1986**, *39*, 337–360.
23. Kora, M.; Schultz, G. Lower Carboniferous Palynomorphs from Um Bogma, Sinai (Egypt). *Grana* **1987**, *26*, 53–66. [CrossRef]
24. Kora, M. Carboniferous Miospore assemblages from the Abu Rodeiyim boreholes, west-central Sinai, Egypt. *Rev. Micropaleontol.* **1993**, *36*, 235–255.
25. El Atfy, H.; Bek, J.; Hosny, A.M.; Uhl, D. The Carboniferous spore genus *Vestispora*: New palynological insights from Gondwana. *Rev. Palaeobot. Palynol.* **2024**, *331*, 105207. [CrossRef]
26. Schrank, E. Paleozoic and Mesozoic palynomorphs from the Foram 1 Well (Western Desert, Egypt). *Neues Jahrb. Geol. Paläontol. Monatshefte* **1984**, *2*, 95–112. [CrossRef]
27. El Ghazaly, G.; Ali, S.M. Middle-Upper Devonian and Lower Carboniferous Miospore Assemblages from Western Desert of Egypt. *Qatar Univ. Sci.* **1985**, *5*, 169–186.
28. Schlumberger. *Well Evaluation Conference, Egypt*; Schlumberger Technical Editing Services, Ltd.: Chester, UK, 1995; 87p.
29. Brown, R.N. History of exploration and discovery of Morgan, Ramadan, and July oilfields, Gulf of Suez, Egypt. In *AAPG Special Publications Facts and Principles of World Petroleum Occurrence—Memoir*; Canadian Society of Petroleum Geologists: Calgary, AB, Canada, 1980; pp. 733–764.
30. Klitzsch, E. Paleogeographical development and correlation of Continental Strata (former Nubian Sandstone) in northeast Africa. *Afr. Earth Sci. (Middle East)* **1990**, *10*, 199–213. [CrossRef]
31. GUPCO. Stratigraphic Log and Location Map of J-62-86 Well. Gulf of Suez, Egypt. 1989, *unpublished manuscript*.
32. GUPCO. Stratigraphic Log and Location Map of J-62-64 AST1 Well, Gulf of Suez, Egypt. 1995, *unpublished manuscript*.
33. Maher, A.; Legrand, J.; Yamada, T.; Komatsu, T. Early Land Plant Spore Assemblage from the Devonian Nakazato Formation of the South Kitakami Belt, Northeast Japan. *Paleontol. Res.* **2024**, *28*, 351–360. [CrossRef]
34. Owens, B.; Streel, M. *Hymenozonotriletes lepidophytus* Kedo, its distribution, and significance about the Devonian-Carboniferous boundary. *Rev. Palaeobot. Palynol.* **1967**, *1*, 141–150. [CrossRef]
35. Clayton, G.; Owens, B.; Al-Hajri, A.; Filatoff, J. Latest Devonian and Early Carboniferous Miospore Assemblages from Saudi Arabia. In *Gulf PetroLink*; Al-Hajri, S., Owens, B., Eds.; GeoArabia Special Publication: London, UK, 2000; Volume 1, pp. 146–153.
36. Loboziak, S.; Melo, J.H.G.; Streel, M. Devonian palynostratigraphy in western Gondwana. In *Applied Stratigraphy*; Eduardo, A.M.K., Ed.; Springer: Dordrecht, The Netherlands, 2005; pp. 73–99.
37. Richardson, J.B.; McGregor, D.C. Silurian and Devonian spore zones of the Old Red Sandstone continent and adjacent regions. *Geol. Surv. Can. Bull.* **1986**, *364*, 1–79.
38. García Muro, V.; Rubinstein, C.V.; Camina, S.C.; Vergani, G.; Pereira, M. Late Early to Middle Devonian Palynology of the Ramos X-1011 Borehole from Northwestern Argentina. 2024. Available online: https://ssrn.com/abstract=4978545 (accessed on 7 October 2024).
39. Obukhovskaya, T. Miospores of the Givetian-Frasnian boundary deposits in Belarus. *Acta Palaeobot.* **2000**, *40*, 17–23.
40. Marshall, J.E.; Reeves, E.J.; Bennett, C.E.; Davies, S.J.; Kearsey, T.I.; Millward, D.; Smithso, T.R.; Browne, M.A. Reinterpreting the age of the uppermost 'Old Red Sandstone' and Early Carboniferous in Scotland. *Earth Environ. Sci. Trans.* **2018**, *109*, 265–278. [CrossRef]
41. Prestianni, C.; Sautois, M.; Denayer, J. Disrupted continental environments around the Devonian-Carboniferous Boundary: Introduction of the timer event. *Geol. Belg.* **2016**, *19*, 135–145. [CrossRef]

42. Clayton, G. Carboniferous miospore and pollen assemblages from the Kingdom of Saudi Arabia. *Rev. Palaeobot. Palynol.* **1995**, *89*, 115–123. [CrossRef]
43. Di Pasquo, M.; Noetinger, S.; Isaacson, P.; Grader, G.; Starck, D.; Morel, E.; Folnagy, H.A. Mid-Late Devonian assemblages of herbaceous lycophytes from northern Argentina and Bolivia: Age assessment with palynomorphs and invertebrates and paleobiogeographic importance. *J. S. Am. Earth Sci.* **2015**, *63*, 70–83. [CrossRef]
44. Di Pasquo, M.; Amenábar, C.R.; Noetinger, S. Middle Devonian microfloras and megafloras from western Argentina and southern Bolivia: Their importance in the palaeobiogeographical and palaeoclimatic evolution of western Gondwana. *Geol. Soc. Lond. Spec. Publ.* **2009**, *314*, 193–213. [CrossRef]
45. Streel, M.; Di Pasquo Lartigue, M. *Quantitative Approach by Miospores of the Devonian-Carboniferous Transition*; NSDS IUGS: Paris, France, 2022; pp. 23–42.
46. Millay, M.A.; Taylor, T.N. Morphological studies of Paleozoic saccate pollen. *Palaeontogr. B* **1974**, *147*, 75–99.
47. Marshall, J.E.; Hemsley, A.R. A Mid Devonian seed-megaspore from East Greenland and the origin of the seed plants. *Palaeontology* **2003**, *46*, 647–670. [CrossRef]
48. Vanguestaine, M. Late Devonian and Carboniferous acritarch stratigraphy and paleogeography. *Ann. Soc. Géol. Belg.* **1986**, *109*, 93–102.
49. Wellman, C.H.; Graham, L.E.; Lewis, L.A. Filamentous green algae from the Early Devonian Rhynie chert. *PalZ* **2019**, *93*, 387–393. [CrossRef]
50. Xiao-Si, F.; Steemans, P.; Streel, M. A new advancement of the Devonian-Carboniferous boundary in Central Hunan Province, China. *Chin. Sci. Bull.* **1993**, *38*, 385–389.
51. Xu, H.H.; Yang, N.; Bai, J.; Wang, Y.; Liu, F.; Ouyang, S. Palynological assemblage of the Lower Devonian of Hezhang, Guizhou, southwestern China. *Rev. Palaeobot. Palynol.* **2022**, *297*, 104561. [CrossRef]
52. Cascales-Miñana, B.; Xue, J.Z.; Rial, G.; Gerrienne, P.; Huang, P.; Steemans, P. Revisiting the spore assemblages from the Lower Devonian Posongchong Formation of Wenshan, Yunnan Province, southwestern China. *Earth Environ. Sci. Trans. R. Soc. Edinb.* **2017**, *108*, 339–354. [CrossRef]
53. Marshall, J.E.; Zhu, H.; Wellman, C.H.; Berry, C.M.; Wang, Y.; Xu, H.; Breuer, P. Provincial Devonian spores from South China, Saudi Arabia and Australia. *Rev. Micropaléontol.* **2017**, *60*, 403–409. [CrossRef]
54. Quetglas, M.; Macluf, C.; Pasquo, M.D. Morphology of the megaspore *Lagenoisporites magnus* (Chi and Hills 1976) Candilier et al. (1982), from the Carboniferous (lower Mississippian: Mid-upper Tournaisian) of Bolivia. *An. Acad. Bras. Ciênc.* **2019**, *91*, e20180750. [CrossRef]
55. Marshall, J.E.; Mangerud, G.; Bringué, M.; Bujak, J. Devonian palynoevents in the circum-Arctic region. *Atl. Geosci.* **2022**, *58*, 307–328. [CrossRef]
56. Potonié, V.R.; Kremp, G. Die Gattungen der paläozoischen Sporae dispersae und ihre Stratigraphie. *Geol. Jahrb.* **1954**, *69*, 111–194.
57. Al-Ghazi, A. New evidence for the Early Devonian age of the Jauf Formation in northern Saudi Arabia. *Rev. Micropaléontol.* **2007**, *50*, 59–72. [CrossRef]
58. Wellman, C.H.; Steemans, P.; Lopes, G. Dispersed spore assemblages from the Lower Devonian Rañeces-La Vid groups of Northern Spain: Palaeogeographical implications. *Rev. Palaeobot. Palynol.* **2023**, *310*, 104825. [CrossRef]
59. Clayton, G.; Coquel, R.; Doubinger, J.; Gueinn, K.J.; Loboziak, S.; Owens, B.; Streel, M. Carboniferous miospores of Western Europe: Illustration and zonation. *Meded.-Rijks Geol. Dienst* **1977**, *29*, 1–71.

life

MDPI

Article

Samotherium boissieri from the Late Miocene of Southern Italy

Antonella Cinzia Marra

Department MIFT, University of Messina, Viale Stagno D'Alcontres 31, I-98166 Messina, Italy; antonella.marra@unime.it; Tel.: +39-090-6765452

Abstract: *Samotherium boissieri* is a giraffid typical of the Pikermian biome, well documented at Samos and occurring in the late Miocene of the Greco-Iranian paleobioprovince. The species has been also recorded at Cessaniti in Calabria (Southern Italy), in a faunal association including other Pikermian species as well as species of Eurasian and African affinity. In this paper, Calabrian specimens are studied and compared to Samos ones. Morphological and biometrical data fall within the variability of *Samotherium boissieri* and clearly differ from the co-occurring giraffid, *Bohlinia attica*. Two partially complete forelimbs, probably referring to the same individual, permit the first full description of the *manus* bones for the species, carpals in particular. The occurrence of *Samotherium boissieri* in Calabria contributes to the wide discussion on paleobiogeographical assessments of the central Mediterranean in the late Miocene, still not well-understood.

Keywords: Giraffidae; Pikermian biome; central Mediterranean; *Samotherium*; *Bohlinia*; *Stegotetrabelodon*; paleogeography; paleoecology

1. Introduction

The late Miocene outcrops of Capo Vaticano–Monte Poro (Calabria, Southern Italy) contain a peculiar mammal fauna, with Greco-Iranian and Afro-Arabian taxa. The mammals come from a transgressive marine succession, well exposed at Cessaniti (the Province of Vibo Valentia, Calabria, Southern Italy; Figure 1). The main site is Gentile's Quarry, near Cessaniti, where the quarrying dismantled huge quantities of sandstones and brought to light (sometimes destroyed) an impressive amount of fossil invertebrates and vertebrates [1,2]. The most iconic fossil of the whole area is the echinoid *Clypeaster*, which can be found very easily and in large quantities [3]. The fossil bones of Sirenian *Metaxytherium serresii* are also abundant, and easily recognized by the pachiostotic structure that, when broken, makes them stand out for the reddish color among the light sandstones [4,5]. Since 1993, terrestrial mammals have been found and identified at Gentile's Quarry and correlatable sites, revealing the presence of Afro-Arabian (*Stegotetrabelodon syrticus* [6–8] and cf. *Certatotherium advenientis* [9]) and Greco-Iranian taxa (*Bohlinia attica* [10,11], *Tragoportax* cf. *rugosifrons* [12], and *Samotherium boissieri* [11]), besides still undetermined Bovids and an antracotheriid [2,10,11,13].

Samotherium boissieri, a giraffid of the Greco-Iranian bioprovince in the late Miocene from about 9–7 Ma, was replaced by its daughter species, *S. major*, characterized by larger dimensions [14]. At Samos, the former is particularly well represented [14,15], to the point that its genus name derives from the site. The occurrence of *Samotherium boissieri* confirms the presence of Greco-Iranian species in the central Mediterranean area, opening the discussion of the possible position and connection of the Cessaniti land during the late Miocene. From approximately 8.7 to 5.3 Ma, the Greco-Iranian bioprovince hosted the Pikermian biome, a characteristic association of large mammals spread in an environment where dense forests alternated with grasslands and shrubs [16–18]. Is the presence of Greco-Iranian species a westward expansion of the Pikermian biome sensu [17,19]? If yes, by what dispersal route, and has the land of Cessaniti had a very different geographical position in the complex geology of the central Mediterranean?

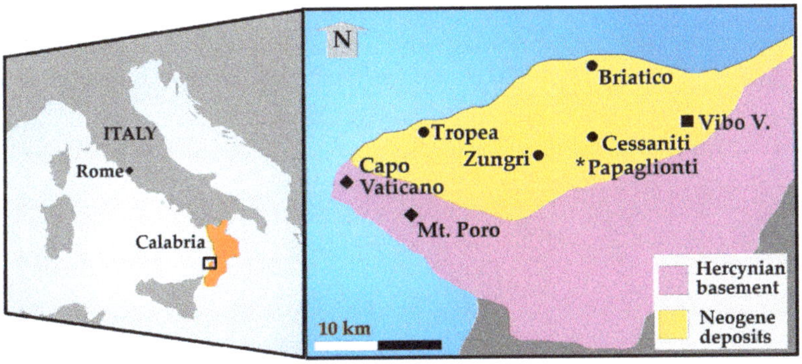

Figure 1. Geographic location of the fossiliferous sites and main geographic references. Square: provincial capital; dots: municipality, town; star: village; diamonds: geomorphological elements.

In this paper, the fossils are attributed to *Samotherium boissieri* by comparison to the wider Samos sample [14]. Moreover, they are compared to *Bohlinia attica*, also occurring in the Cessaniti mammal assemblage [10].

In addition to the scientific aspect, the occurrence of *Samotherium boissieri* in Calabria recalls an ante litteram travel from Greece to Magna Graecia. Curiously, the species is typical of Samos, one of the places where later would flourish the Ancient Greek civilization, and occurs in Calabria, part of the Magna Graecia, a wide area settled by Greek colonists from the 8th century B.C. Even more curiously, a skull of *Samotherium* seems to be figured in a Greek vase as a monster and traced back to ancient finds of fossils [19,20]. The so-called "Monster of Troy" figured on a VI century BC Corinthian vase has been identified as a *Samotherium* by the proportions of the skull with respect to humans and for the short anterior maxilla, so figured for the large nasal cohanes of the giraffid [19,20]. The "monster of Troy" has been seen as a monitor lizard by Monge-Nájera [21]. The link among myth, legend, and fossil is intriguing, and has numerous points of evidence also at Samos, where ancient Greeks identified fossil mammals as Neades or Amazons [22]. The discovery of bones resembling those of domestic animals, but with "monstrous" sizes and proportions, along with the occurrence of earthquakes, led the inhabitants of Samos to attribute them to the Neades, horrible beasts that had ravaged the island and emitted sounds so loud that they caused cracks in the ground [22]. The discovery of fossil bones, potentially including a skeletal element of *Samotherium*, in the temple of Hera in Samos is particularly interesting, as they were likely regarded as relics of the Neades [22]. The discovery of bones in the reddish sediments of Panaima at Samos has been attributed to a battle of the Amazons, who fell with their horses, likely recognized in the fossils of *Hipparion* [22]. At Capo Vaticano–Monte Poro, direct connections between fossils and myth are not documented. The fossils of *Clypeaster* enter the local folklore because they are used in fragments to make fire crackling in fireplaces. However, a local myth indicates the Greek polis of Hipparion (the current city of Vibo Valentia, Figure 1) as a place chosen by the Gods for the enchanting landscape. Due to this legend, the coast from Tropea to Briatico (Figure 1) is still named "Coast of the Gods".

Localities and Geological Setting

The studied fossils come from an abandoned quarry at Malopara (the municipality of Zungri, the Province of Vibo Valentia, Calabria, Southern Italy; Figure 1) and from Gentile's Quarry (the municipality of Cessaniti, the Province of Vibo Valentia, Calabria, Southern Italy; Figures 1–3). The sites are included in the Capo Vaticano–Monte Poro area, where a Neogene succession rests on a Paleozoic crystalline substratum from the inland to the coast [1,2,4,23] (Figures 1 and 2).

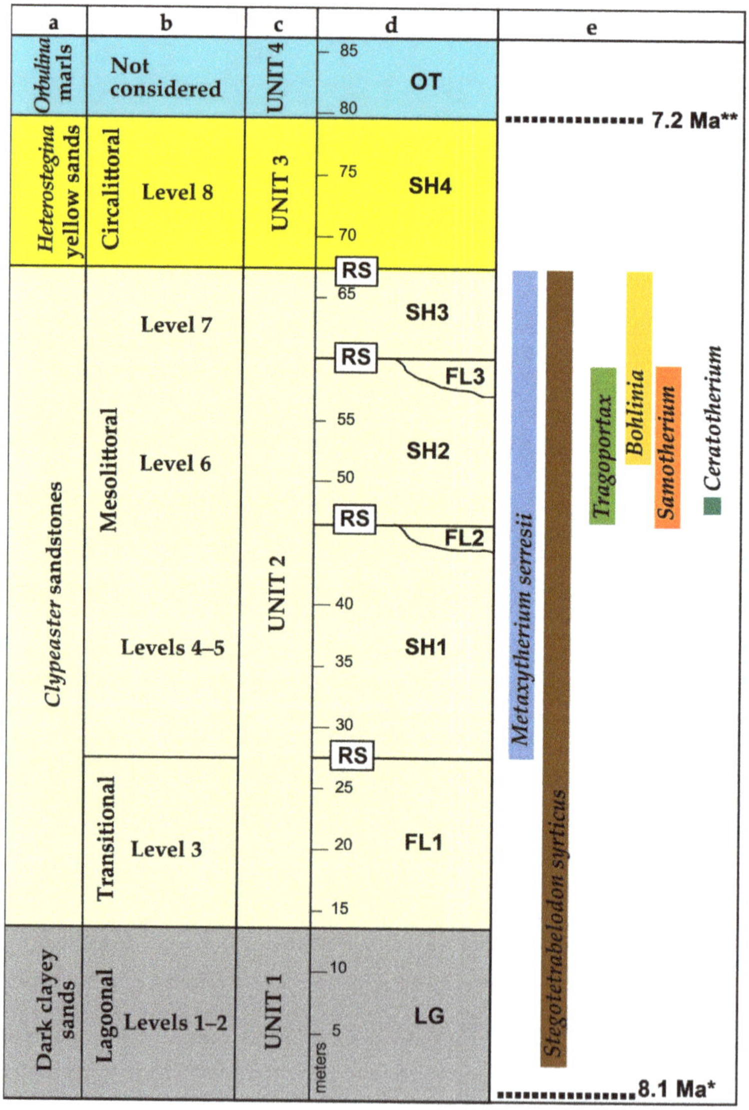

Figure 2. Stratigraphy of Gentile's Quarry outcrop according to: column a, Nicotera [24]; column b, Carone and Domning [4]; column c, Gramigna [1]; column d, Marra et al. [2]. Column e: occurrences of mammals and dating based on magnetostratigraphy (*) and nannoplankton biozones (**) according to Marra et al. [2]. LG: lagoonal deposits; FL: fluvial deposits; SH: shoreface deposits; OT: offshore transition; RS: ravinement surface.

(a) (b)

Figure 3. The abandoned quarry at Malopara: (**a**) overview; (**b**) level with *Samotherium*.

The most representative succession of the area is well exposed at Gentile's Quarry (Cessaniti), which produced the main part of the fossil mammals, bought to light by quarrying works and collected by amateur paleontologists [24]. The succession, schematized in Figure 2 according to different authors [1,2,23,25], represents a marine transgression, with some intervals represented by fluvial deposits (FL in Figure 2) followed by ravinement surfaces (RS in Figure 2; intended as erosional surfaces formed during transgressions), probably occurring under tectonic control. In proximity to the ravinement surfaces, bone beds of *Metaxytherium* and other mammals are found [1]. The succession has an age between 8.1 and 7.2 Ma [2]. In about 1 Ma, the transgression changed the environmental conditions at Cessaniti, from marginal lagoons protected by sandy barriers to a shallow sea up to marine conditions [1,2,26–28].

The specimens studied in this paper come from the upper part of the "*Clypeaster* sandstones" outcropping throughout the area of Cessaniti–Zungri with different thickness and characterized by the impressive quantity of echinoid shells and sirenian bones, attributed to *Metaxytherium serresii*. From the *Clypeaster* sandstones come a rich land mammal assemblage (Figure 2, column e), including *Bohlinia attica* [2,10,11], *Tragoportax* cf. *rugosifrons* [12], *Stegotetrabelodon syrticus* [6–8], cf. *Ceratotherium advenientis* [9], still undetermined Bovids and Anthracotheriids [2,11,13], and *Samotherium boissieri* studied in this paper.

The taphonomy of the site is not fully clear because the fossils have not been collected through systematic excavation (see Section 2, Materials and Methods). A large part of mammal remains present abrasion and damage by river transport, while many also show incrustations by marine organisms. It is quite evident that the mammal remains reached the sea carried by rivers to the seabed before final burial, providing a substratum to encrusting marine organisms. At Malopara (Zungri; Figures 1 and 3), two forelimbs, probably belonging to the same individual of *S. boissieri*, have been found from levels correlatable to upper SH1-SH2 (Figure 2 column d) of the *Clypeaster* sandstones. As reported by the collectors, the bones were slightly scattered at close range in two groups: one including left radius, carpals, and metacarpal; one including some right carpals, sesamoids, metacarpal, and phalanges. A fragmentary humerus is too damaged to be included in this study. Some bones are altered by weathering; many are slightly abraded and some present encrustations by marine organisms. It can be assumed that the carcass was transported by a river still with connective tissues, maybe during a flood event, and then laid down on the seabed in partial anatomical connection. There is also the possibility that the animal died near the sea. Research to find other skeletal elements has not been successful. The find is unique for its completeness.

At Gentile's Quarry, two metacarpals and an astragalus of *S. boissieri* come from the level SH2 of the *Clypeaster* sandstones.

2. Materials and Methods

The materials studied are in Table 1, with the locality and the inventory number attributed by the institution where the fossils are exhibited [24,29]: MuRi is for the City Museum of Ricadi (Vibo Valentia) and SiMU is for the Museal System of the University of Calabria (Cosenza). An astragalus without inventory, studied in 2010, is included in this paper. The astragalus was collected in the "*Clypeaster* sands" at Cava Gentile by the late Mario Bagnato and stored at Parghelia (VV) in the headquarters of the "Gruppo Paleontologico Tropeano" (GPT), a former amateur association. The hall and storage rooms were later integrated into the "Museo della Memoria" (MuMe) in Parghelia; however, the astragalus is currently unavailable, and an inventory check has been requested.

Table 1. Giraffidae from Cessaniti described and compared in this paper.

Specimen	Locality	Inventory Number
Samotherium boissieri		
Ulna—left proximal epiphysis	Malopara	MuRi-870
Radius—left distal epiphysis	Malopara	MuRi-895
Radius—right distal epiphysis	Malopara	MuRi-866
Scaphoid—left	Malopara	MuRi-878
Intermedium—left	Malopara	MuRi-879
Intermedium—right	Malopara	MuRi-885
Ulnare—left	Malopara	MuRi-880
Ulnare—right	Malopara	MuRi-884
Magnotrapezoid—left	Malopara	MuRi-882
Unicinatum—left	Malopara	MuRi-881
Unicinatum—right	Malopara	MuRi-886
Sesamoids and pisiform	Malopara	MuRi-883
Metacarpal—left	Malopara	MuRi-865
Metacarpal—right	Malopara	MuRi-864
Metacarpal. distal end—left	Cessaniti	MuRi-875
First phalanx, III digit—right	Malopara	MuRi-888
First phalanx, IV digit—right	Malopara	MuRi-887
Second phalanx, III digit—right	Malopara	MuRi-890
Second phalanx, IV digit—right	Malopara	MuRi-899
Third phalanx, III digit—right	Malopara	MuRi-892
Third phalanx, IV digit—right	Malopara	MuRi-891
Third phalanges III and IV digit	Cessaniti ?	MuRi-894
Astragalus—right	Cessaniti	? GPT-MuMe
Metatarsal—left	Cessaniti	SiMU-CMA07
Metatarsal—right	Cessaniti	SiMU-CMA06
Bohlinia attica		
Radius—left	Cessaniti	MuRi-874
Metacarpal—left	Cessaniti	MuRi-972
Astragalus—left	Cessaniti	SiMU-CMA02
Astragalus—right	Cessaniti	Muri-877

Uncertain provenance or storage are marked with "?".

The measurements have been taken with a manual caliper, following the standards of Von Den Driesch [30] adapted for the definition of Kostopoulos [14] to render clear the comparisons. The RI (Robustness Index) is indicative of metapodial slenderness and is calculated as the transverse diameter of the shaft/the length of the shaft.

The fossils have been compared to the other occurring giraffid, *Bohlinia attica* (Table 1). The anatomical terminology adopted is from Bohlin [31], Kostopoulos [32,33] Rios et al. [34], and Solounias and Danowitz [35].

The collecting methods have not been systematic for the quarrying and for the great dispersion of the fossils. Fossils were collected through the ongoing efforts of amateur paleontologists, which gathered emerging fossils during breaks in quarrying activities at Gentile's Quarry and nearby areas, particularly after rainfall, noting the site and altitude of each find, though not always adhering to strict scientific protocols [24]. The collectors referenced Nicotera's informal stratigraphy (column "a" in Figure 2, [23]) and recorded specific conditions useful to track the finds in the stratigraphic column [24]. This approach allowed them to retrieve many fossils that would otherwise be destroyed and that would not be systematically excavated due to their low concentration in the sediment. This collection method yielded a rich fossil record [24,29], but compromised the taphonomic analysis.

3. Systematic Paleontology

Class: Mammalia Linnaeus, 1758;
Order: Artiodactyla Owen, 1848;
Suborder: Ruminantia Scopoli, 1777;
Family: Giraffidae Gray, 1821;
Genus: *Samotherium* Forsyth-Major, 1888;
Species: *Samotherium boissieri* Forsyth-Major, 1888;
Studied materials are listed in Table 1.

Stratigraphic occurrence: Late Miocene (Upper Tortonian; European Land Mammal Age: Turolian) of Cessaniti and Zungri (Vibo Valentia, Italy), *Clypeaster* sandstones (levels 6 and 7 according to Carone and Domning [4], corresponding to SH2-SH3 in Marra et al.'s study [2]).

3.1. Descriptions

Forelimbs

Two partial forelimbs attributable to the same individual come from Malopara. The left forelimb includes distal radius (MuRi-895), proximal ulna (MuRi-870; Figure 4a), all carpals (MuRi878, 880,879,881,882,883), and metacarpal (MuRi-865). The right forelimb includes some carpals (MuRi 884, 885, 886), metacarpal (MuRi-864), first phalanx of the III toe (MuRi-888), first phalanx of the IV digit (MuRi-887), second phalanx of the III digit (MuRi-890), second phalanx of the IV digit (MuRi-889), third phalanx of the III digit (MuRi-892), and third phalanx of the IV digit (MuRi-891).

In the preserved proximal ulna (MuRi-870, Figure 4a), the posterior margin of the olecranon is straight. The ulnar notch is curved and wide. Measurements: depth across the anconeal process = 100.4 mm; smallest breadth of the olecranon = 85; breadth across the coronoid process = 56.4.

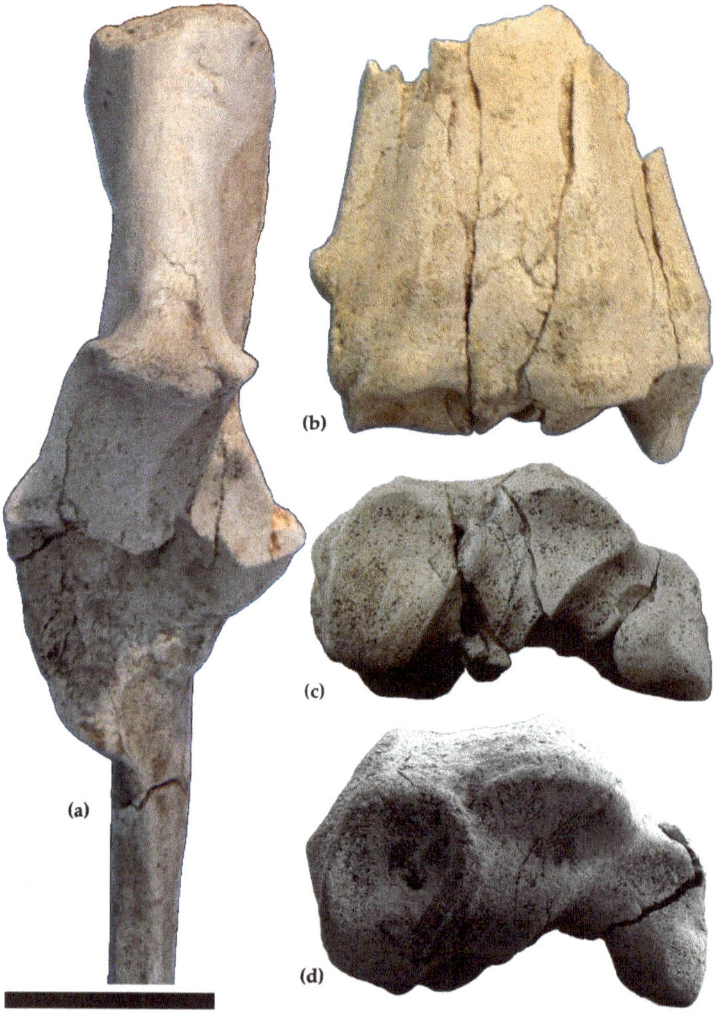

Figure 4. *Samotherium boissieri* from Malopara: left proximal ulna (MuRi-870) in frontal view (**a**); left distal radius (MuRi 895) in dorsal (**b**) and distal (**c**) views; *Bohlinia attica* from Cessaniti: (**d**) left distal radius in distal view (MuRi-874). Scale bar: 5 cm.

The two distal radii (MuRi-866 and MuRi-895, Figure 4b,c) are fragmented. The styloid process is prominent. The groove for the extensor carpi radialis and the groove for the common extensor tendon are well evident. The ulnar notch is prominent. Despite fragmentation, it is possible to describe the distal articular surface, which presents oblique facets for the scaphoid and the intermedium, and a deep facet for the ulnare is deep, subvertical in the margin with the facet for the intermedium. The fragmentation of the bones makes reliable measurements difficult; the approximate distal breadth of MuRi-895 is 96.4 mm.

Carpal bones belonging to the same individual of *Samotherium boissieri* are here described in their entirety for the first time. In the absence of comparison specimens for most of them, they can be attributed to the species by association with the metacarpals, which clearly have the characters of *S. boissieri*. The bones are figured in anatomical connection (Figure 5), but were found slightly scattered.

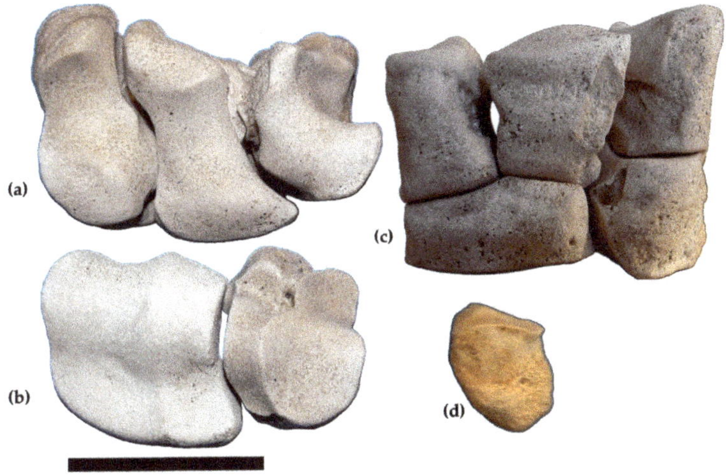

Figure 5. *Samotherium boissieri* from Malopara. Carpals in anatomical connection: (**a**) proximal row of left carpals in proximal view (scaphoid MuRi-878, intermedium MuRi-879 and ulnare MuRi-880); (**b**) distal row of left carpals in proximal view (magnotrapezoid MuRi-882, uncinatum MuRi-881); (**c**) carpals in anatomical connection in anterior view; (**d**) pisiform MuRi-883. Scale bar: 5 cm.

Scaphoid (MuRi-878, left, Figure 5a,c). Proximal view: the articular surface enlarges towards the anterior side; it is high and convex posteriorly, proceeds uniformly concave, and drops down anteriorly. Lateral view: the proximal articular surface continues laterally, where forms an anteroposterior elongated facet for the semilunar; on the lateral side a quadrangular facet for the semilunar is present distally. Medial view: the bone is flat and rough, without facets. Distal view: the facet for the magnotrapezoid is irregularly rectangular. The posterior side of the bone presents a protruding tuberosity. Greatest breadth (MuRi-878): 30.6 mm.

Intermedium (MuRi-879, left, Figure 5a,c; MuRi-885, right). Proximal view: the articular surface is narrow, convex and elevated posteriorly, narrower in the middle, where it is concave, and convex, elevated and wide anteriorly. Lateral view: an irregularly squared facet is proximally for the ulnare; a sub-oval facet extends upwards from the distal articular surface, in a central-posterior position. Medial view: in the upper part, an elongated facet is present for the scaphoid; another facet for the scaphoid is located anteriorly, near the distal margin. Distal view: the articular surface for the magnotrapezoid is convex anteriorly and concave posteriorly. Greatest breadth of MuRi-879: 31 mm; MuRi-885: 32 mm.

Ulnare (MuRi-880, left, Figure 5a,c; MuRi-884, right). Proximal view: the articular surface is high and convex posteriorly and drops down anteriorly, where it is concave; it is separated by a rim in a wider medial portion and a narrower lateral portion. Lateral view: two facets for the uncinatum are present, a quadrangular one anteriorly and a semicircular one posteriorly. Medial view: the bone is rough and has no facet. Distal view: the articular surface for the metacarpal is flat, quadrangular, and wider anteriorly. Greatest breadth of MuRi-880: 30 mm; MuRi-884: 29 mm.

Magnotrapezoid (Muri-882, left, Figure 5b,c). Proximal view: the articular surface is high and convex posteriorly and drops down anteriorly, where it is concave; it is separated by an antero-posterior rim in a wider medial articulation and a narrower lateral articulation. Lateral view: the articulation for the uncinatum consists of an anterior quadrangular facet and a posterior semicircular facet. In the medial view, the bone is flat, without facets. Distal view: a flat and quadrangular facet for the metacarpal is present. Greatest breadth of MuRi-882: 42 mm.

Uncinatum (Muri-881, left, Figure 5b,c; Muri-886, right). Proximal view: the articular surface is higher and larger anteriorly and extends in the posterior side; it is separated by a rim in a medial convex facet and a lateral concave facet. Lateral view: facets not present. Medial view: a semicircular articular facet extends around a shallow depression. Posterior view: the upper part is occupied by the extension of the proximal surface. Distal view: a flat articular surface is for the metacarpal. Greatest breadth of MuRi-881: 36.7; MuRi-886: 36.6.

Pisiform. The pisiform (MuRi-883, left; Figure 5d) is robust with an elongated facet for the ulnare. Greatest breadth: 27.7.

The left (MuRi-865, Figure 6a–c) and the right (MuRi-864, Figure 6d,e) metacarpals have a slender diaphysis, with a RI = 11.8 (MuRi-864) and 11.7 (MuRi-865). The synovial fossa is closed. The proximal articulation has a semicircular profile, and the postero-medial and palmar tubercles are absent. The proximal articular surface has a flat rectangular facet for the capitatum, more elongated in the lateral direction. It is separated by a step from the flat squared facet for the uncinatum, lying on a lower plane. In the posterior view, the shaft has a shallow sulcus, delimitated by wide and smoothed crests.

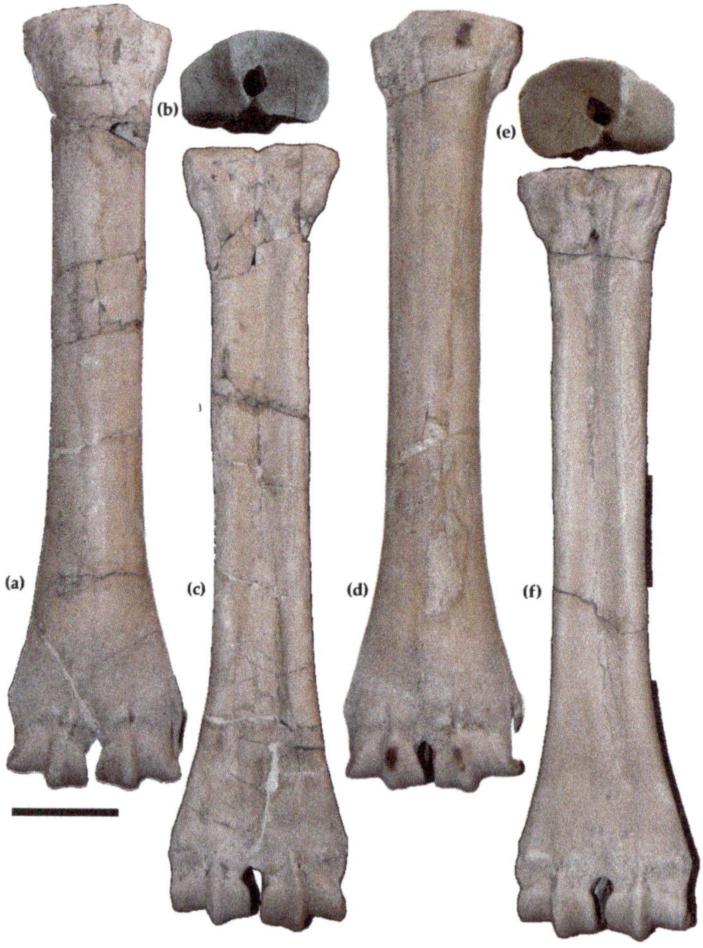

Figure 6. *Samotherium boissieri* from Malopara. Left metacarpal (MuRi-865) in dorsal (**a**), proximal (**b**), and ventral (**c**) views. Right metacarpal (MuRi-864) in dorsal (**d**), proximal (**e**), and ventral (**f**) views. Scale bar: 5 cm.

The measurements of the metacarpals (MuRi-864, MuRi 865, and MuRi-875, Table 2) fall in the variability of *S. boissieri*, and are sensibly smaller than *Bohlinia* from Cessaniti (MuRi-972, Table 2).

Table 2. Measurements of the metacarpals. L: length; TDp: transverse diameter of proximal epiphysis; APDp: antero-posterior diameter of proximal epiphysis; TDdia: transverse diameter of diaphysis; CD: circumference of diaphysis; APDdia: antero-posterior diameter of diaphysis; TDd: transverse diameter of distal epiphysis; APDd: antero-posterior diameter of distal epiphysis.

Inventory	L	TDp	APDp	TDdia	CD	APDdia	TDd	APDd
MuRi-865	373	75.7	48.2	43.7	136.5	36.5	84.1	48.2
MuRi-864	373	74.4	49.2	44	136.5	36.3	82.8	48.0
MuRi-875	-	-	-	-	-	-	83.4	48.0
MuRi-972	-	90.8	53.1	-	-	-	88.0	49.9

The sesamoids are characterized by elongated articular facets.

The proximal phalanges (MuRi-888, MuRi-887; Figure 7a,b; Table 3) belonging to the right manus are slender, larger proximally than distally. The intermediate phalanges are short (MuRi-880, MuRi-889; Figure 7a; Table 3). The distal phalanges (MuRi-892, MuRi-891, Figure 7a; Table 3) have very concave facets; the external one is larger than the internal one.

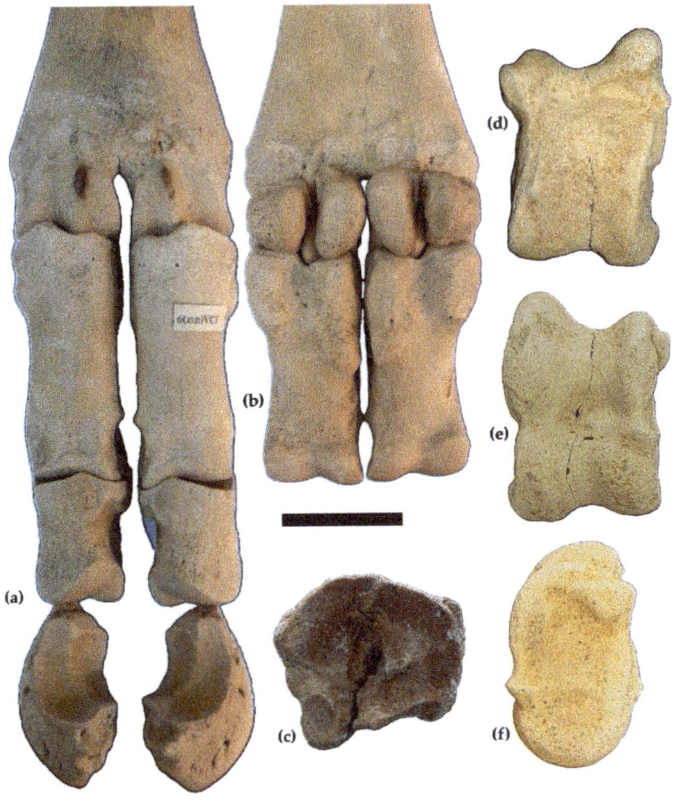

Figure 7. *Samotherium boissieri*. Right phalanges (MuRi-888, 887, 890, 889, 892, 891) in anatomical connection with the metacarpal (MuRi-864) in dorsal view (**a**) and detail of the ventral view with sesamoids (**b**), from Malopara. Left metatarsal (SiMU-CMA07) in proximal view (**c**); right astragalus (GPT-MuMe) in ventral (**d**), dorsal (**e**), and medial (**f**) views; from Cessaniti. Scale bar: 5 cm.

Table 3. Measurements of the phalanges of *Samotherium boissieri* from Zungri. L: length; TDp: transverse diameter of proximal epiphysis; SD: smallest breadth of diaphysis; TDd: transverse diameter of distal epiphysis; DLS: greatest diagonal length of the sole; Ld: length of dorsal surface; MBs: middle breadth of the sole.

Specimen	Inventory	L	TDp	SD	TDd
First phalanx, III digit	MuRi-888	100.9	41.1	32.9	32.2
First phalanx, IV digit	MuRi-887	98.0	40.1	32.7	31.4
Second phalanx, III digit	MuRi-890	65.2	37.7	30.6	32.2
Second phalanx, IV digit	MuRi-899	61.3	36.9	30.9	31.0
		DLS	**Ld**	**MBs**	
Third phalanx, III digit	MuRi-892	80.7	59.1	30.5	
Third phalanx, IV digit	MuRi-891	80.5	61.7	29.3	

The astragalus (GPT-MuMe, Figure 7d–f) is massive. Dorsal view: the lateral ridge of the trochlea is more elevated and slightly larger than the medial one; the medial ridge of the trochlea protrudes with a tuberosity toward the medial side; the groove of the trochlea is U-shaped; the central fossa is shallow; the head has a lateral articulation elongated in antero-posterior direction; the medial bulge of the collum talii is located almost at 1/3 of the medial margin. Ventral view: the proximal triangular fossa is deep and laterally extended until the interarticular groove; the medial scala is well evident. Medial length: 80.0 mm; lateral length: 90 mm; depth of lateral half: 48; depth of medial half: 54; distal transverse diameter: 58.

The metatarsals (SiMU-CMA06 and SiMU-CMA07, Table 4; Figure 7c) have a robust shaft furrowed by a wide posterior groove delimited by two crests located near the medial and the lateral sides. In the proximal epiphysis, the facet for the naviculo-cuboid is slightly convex towards the center, flat externally. The lateral facet for the cuneiform is C-shaped, wide, and flat, slightly convex in its posterior portion. The medial facet for the cuneiform is rectangular with rounded corners.

Table 4. Measurements of the metatarsals of *Samoherium boissieri* from Cessaniti. L: length; TDp: transverse diameter of proximal epiphysis; APDp: antero-posterior diameter of proximal epiphysis; TDdia: transverse diameter of diaphysis; CD: circumference of diaphysis; APDdia: antero-posterior diameter of diaphysis; TDd: transverse diameter of distal epiphysis; APDd: antero-posterior diameter of distal epiphysis.

Inventory	L	TDp	APDp	TDdia	CD	APDdia	TDd	APDd
SiMU-CMA06	395	69.9	65.9	41.0	15.0	33.6	75.8	46.0
SiMU-CMA07	405.0	78.0	72.0	43.0	15.5	37.0	91.3	50.3

3.2. Comparisons

The comparisons are mainly addressed to the large sample of *Samotherium boissieri* from Samos, Greece, as well as to comparative studies and descriptions [14,15,34–38]. Radii, metacarpals, and astragalus are also compared to the specimens of *Bohlinia attica* from Cessaniti [10].

The two forelimbs found at Malopara include radii and metacarpal bones; the latter are very significant for taxonomy and can be considered crucial to attributing the two *mani* found at Malopara.

The two distal radii (MuRi-895, Figure 4b,c, and MuRi-866) are fragmentary and can be attributed to *S. boissieri* for the general morphology and the available measurements. Compared to the radius of *B. attica* from Cessaniti (Muri-874), the distal articular surface of

MuRi-895 has shallower facets for the scaphoid and the intermedium and a deeper facet for the ulnare.

In the metacarpals (MuRi-864, MuRi-965; Figure 6), the semicircular profile and the squared facets of the distal articular surface as well as the shallow posterior sulcus of diaphysis are similar to *S. boissieri* [14,15,34], but differ from the semioval profile of the proximal articular epiphysis, the oval proximal articular facets, and the deep posterior sulcus observed in the right metacarpal of *B. attica* from Cessaniti (Mu.Ri-972; Figure 6; [10]). The measurements and the bone proportions calculated in the dispersion plot (Figure 8) fall within the variability known for *Samotherium boissieri* [14]. The two metacarpals probably belonged to male individuals, by comparison with the range of sexual bimodality [14]. The robustness indices (RI: Transverse diameter of shaft/length; 11.8 in MuRi-864, 11.7 in MuRi-865) fall within the variability known for *Samotherium boissieri* from the Mytilinii Basin of Samos (from 10.4 to 13.6, mean: 12.45, calculated on 12 specimens; [14]) and differs from the variability known for *B. attica* (5.56–8.57; mean value 7.3 calculated on 11 specimens; [34,39–41]).

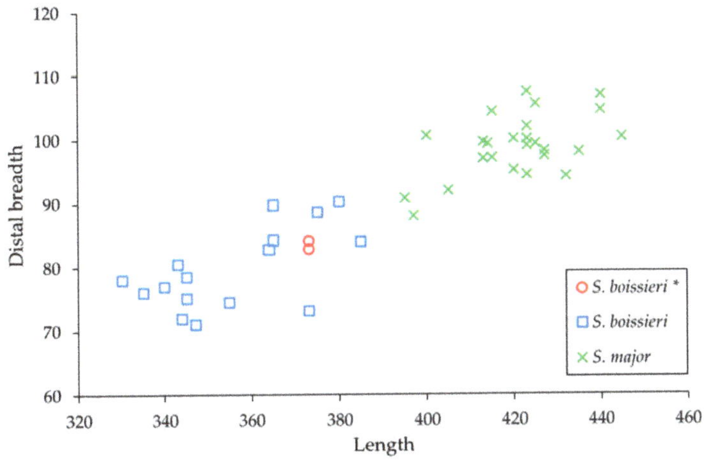

Figure 8. Dispersion plot of the length/distal breadth (transverse diameter) of the metacarpals of *Samotherium boissieri* (squares) and *Samotherium major* (crosses) from Samos [14] and *Samotherium boissieri* from Malopara (circles, marked with *).

The astragalus (GPT-MuMe) falls within the biometrical variability of *Samotherium boissieri* (Figures 7d–f and 9; Table 5). It is worth noting that *Bohlinia attica* from Cessaniti and other sites [35–37,39] also has similar measurements. The marked differences between *Samotherium boissieri* and *Bohlinia attica* astragali are in the morphology, as also observable in the specimens from Cessaniti referred to the two species: the lateral ridge of the throclea is less elevated and robust in *Samotherium* than in *Bohlinia*; the groove of the trochlea is more marked than in *Bohlinia*, where it is wide; the medial bulge of the *collum talii* is located more posteriorly in *Bohlinia*; in antero-posterior direction, the head is larger in *Samotherium* than in *Bohlinia*; the proximal triangular fossa is less extended laterally in *Bohlinia*; the medial scala is more evident in *Samotherium*. Kostopoulos [14] indicated sexual dimorphism ranges expressed by a bimodality also observable in Figure 9; the astragalus from Cessaniti falls within male variability.

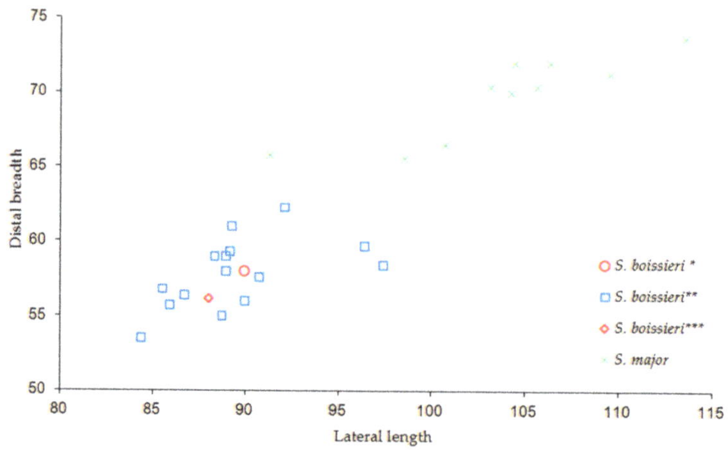

Figure 9. Dispersion plot of the lateral length/distal breadth (transverse diameter) of the astragalus of *Samotherium boissieri* (squares; marked with two *) and *Samotherium major* (crosses) from Samos (measurements after K 2009 [14]), *Samotherium boissieri* (diamond; marked with three *) from Kavakdere [37], and *Samotherium boissieri* from Malopara (circles; marked with one *).

Table 5. Measurements of astragalus of *Samotherium boissieri* and *Bohlinia attica*. Llat: lateral length; Lmed: Medial length; TD: distal transverse diameter; APDd: antero-posterior diameter of distal epiphysis.

Species	Ref.	Llat	Lmed	TD	APDd
Cessaniti specimen		90	80	58	54
Samotherium boissieri					
	[14]	90.08	78.5	57.6	
	[14]	96.5	82.3	59.7	
	[14]	97.5	85.1	58.4	
	[14]	89.2	78.6	59.3	
	[14]	89.0	79.2	58.0	
	[14]	88.4	78.0	59	
	[14]	86.8	73.7	56.4	
	[14]	85.6	75.0	56.8	
	[14]	89.3	76.2	61.0	
	[14]	84.4	76.5	53.5	
	[14]	88.8	77.6	55.0	
	[14]	89.0	75.8	59.0	
	[14]	86.0	77.2	55.7	
	[14]	90.0	76.6	56.0	
	[14]	92.2	81.0	62.3	
	[37]	88.1	80.95	56.16	
Bohlinia attica					
	[10]	89	78.6	59	55.6
	[10]		81	59	
	[36]	87.7–103.4 (min–max)	76.5–89.6 (min–max)	55–68 (min–max)	46.5–61.1 (min–max)
	[36]	103	99	76.6	
	[36]	105	94	70	
	[36]	105	91	71	
	[36]	108	94	73	
	[42]	83.8			59.3
	[42]		87.6		
	[39]		81	59	

The two metatarsals fall within the morphological and biometrical variability of the species *Samotherium boissieri* (Figures 7c and 10) and are considerably shorter than *Bohlinia attica* [39,40]. The robustness indices (RIs) are 10.6 and 10.4, within the variability of *Samotherium boissieri* from Samos (9.5–12.5; mean = 10.7 calculated in 20 specimens [14]). According to the sexual bimodality indicated by Kostopoulos [14], the specimen SiMu-CMA06 falls next to the female size range, while SiMu-CMA06 falls near the male range.

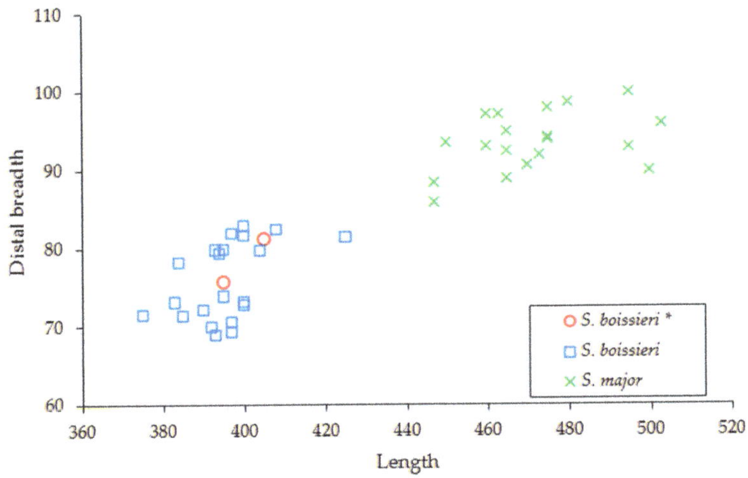

Figure 10. Dispersion plot of the length/distal breadth (transverse diameter) of the metatarsals of *Samotherium boissieri* (squares) and *Samotherium major* (crosses) from Samos [14] and *Samotherium boissieri* from Malopara (circles; marked with *).

4. Discussion

Samotherium boissieri is a species spread across the Greco-Iranian bioprovince, from 8.9 to about 7 Ma (https://nowdatabase.org/ accessed on 4 May 2025); its occurrence at Cessaniti needs broad discussion on biochronology, paleoecology and paleogeography of the whole mammal association, represented by a significant number of fossils (Figure 2, Tables 6 and 7).

Table 6. List of fossils of the land mammals of Cessaniti from Gentile's Quarry and correlatable deposits at Papaglionti (*) and Malopara (**).

	Stegotetrabelodon syrticus	**cf. *Ceratotherium advenientis***	***Tragoportax* cf. *rugosifrons***	***Bohlinia attica***	***Samotherium boissieri***
Unit 2	mandible; incisor; fragmentary molar; two fragmentary humeri *; right II metacarpal; incomplete femur	fragmentary skull; two fragmentary teeth; few postcranial bones	hemimandible; radius; astragalus; anterior phalanx; humerus *; anteriorphalanx *; metatarsal *; posterior phalanx *	upper molar row distal radius; metacarpal; two astragali; two cubonaviculars;	two metacarpals two metatarsals complete manus ** fragmentary ulna ** distal radius ** fragmentary humerus **
Unit 1	one worn DP4				

Table 7. Distribution of *Samotherium boissieri*, *Bohlinia attica*, and *Tragoportax rugosifrons*, data from https://nowdatabase.org/now/database/ by The NOW Community/CC BY 4.0, accessed 4 May 2025; *: several paleontological sites in the locality.

Locality	Country	Age (Ma)	*Samotherium boissieri*	*Bohlinia attica*	*Tragoportax rugosifrons*
Azmaka	Bulgaria	7.2		X	
Gorna Susica	Bulgaria	8.3–7.285		X	
Hadjidimovo	Bulgaria	7.6–7.1		X	X
Kalimantsi	Bulgaria	7.6–7.1		X	
Krodimovo	Bulgaria	7.6–7.1		X	
Strumyani	Bulgaria	7.6–7.1		X	X
Kirokuçuk	N Macedonia	8.9–5.3		X	
Dolni Disan	N Macedonia	8.9–5.3		X	
Karaslari	N Macedonia	7.6–7.1		X	X
Ditiko	Greece	7.1–5.3		X	
Kerassia	Greece	8.9–7.1		X	
Nikiti	Greece	9.9–8.9		X	
Pikermi *	Greece	7.454–7.1		X	
Pyrgos Vassilissis	Greece	7.212–7.14		X	
Ravin de la Pluie	Greece	9.426–9.311		X	
Ravin Zouaves *	Greece	9.9–8.1		X	
Samos *	Greece	8.9–5.3	X		X
Vathylakkos *	Greece	7.489–7.454		X	
Corakyerler	Turkey	8.9–7.6	X	X	
Esendere	Turkey	8.9–7.6		X	
Küçükçekmece	Turkey	8.9–5.3		X	
Gülpinar	Turkey	7.6–7.1	X		
Karain	Turkey	8.9–7.1	X		
Kavakdere	Turkey	8.254–8.108	X		
Sinap *	Turkey	11.2–7.1		X	
Injana	Iraq	8.9–7.6	X	X	
Maragheh *	Iran	8.9–7.1		X	X
Novo-Elizavetovka	Ukraine	8–9–7.1	X		
Taraklia	Moldova	7.6–7.1	X		X

4.1. Biochonology and Distribution

The Artiodactyls occurring at Cessaniti are *Samotherium boissieri*, *Bohlinia attica*, and *Tragoportax* cf. *rugosifrons*. The three species are associated only at Cessaniti, while *Samotherium boissieri* co-occurs with *Bohlinia attica* at Corakyerler (Turkey) and Injana (Iraq) and with *Tragoportax rugosifrons* at Samos (Greece) and Taraklia (Moldova); *Bohlinia attica* and *Tragoportax rugosifrons* co-occur in the Bulgarian sites of Hadjidimovo and Strumyani, and at Karaslari in the Republic of Macedonia (data from https://nowdatabase.org/now/

database/, accessed on 3 May 2025; Table 7). The biochronologic distribution of the three species in the Greco-Iranian bioprovince is coherent with the age of Cessaniti (Table 7).

The species *Stegotetrabelodon syrticus* is only known from the site of Sahabi, dated 6.7 Ma [43] (https://nowdatabase.org/ accessed 3 May 2025). However, until few years ago, the species included Arabian specimens, recently enriched by new finds and attributed to *Stegotetrabelodon emiratus*, bracketed in a time interval of 8–6 Ma [44,45]. The studies on *Stegotetrabelodon* from Cessaniti, carried out in 1993 [6] and 2016 [7], considered the whole variability of *Stegotetrabelodon syrticus*, including the Arabian specimens now called *Stegotetrabelodon emiratus*. Some plesiomorphic characters were inferred on the sample from Cessaniti. A revision of the proboscideans from Cessaniti, also including recent new finds, is in progress. Giraffidae are signaled in the Baynunah Formation (United Arab Emirates): some are indetermined, other are dubiously considered *Palaeotragus*/*Samotherium* [42,46], and a partial skeleton is attributed to *Palaeotragus* aff. *germaini* [47].

The new rhinocerotid, cf. *Ceratotherium advenientis*, is not biochronologically informative given its uncertain systematic position, probably next to the African genera *Diceros* and *Ceratotherium* [9].

4.2. Paleoecology

Samotherium boissieri and *Bohlinia attica* are both considered browsers by Solounias et al. [17,48]. Rios et al. consider *Samotherium boissieri* a seasonal mixed feeder and *Bohlinia attica* a grazer [49]. *Tragoportax rugosifrons* was a grazer and the genus *Tragoportax* probably occupied an ecological niche like that of the modern *Hippotragus*, consisting of forest–savannah environments interspersed with open spaces [17,50]. *Stegotetrabelodon* is considered a ground-dwelling grazer–browser and *Ceratotherium* a ground-dwelling grazer. These palaeoecological characteristics fall in the variability of the Pikermian biome, typical of the Greco-Iranian bioprovince and described as a sclerophyllous evergreen woodland interspersed with grassy meadows, inhabited by a uniform mammal fauna [16,17,51–53]. The uniformity of the Pikermian biome led to the consideration of the fauna as a "chronofauna" in the sense of Olson [51]. However, Kostopoulos [18] challenged this notion, pointing out variations in species diversity and biochronology across different sites. In contrast, Eronen et al. [54] applied the Genus-level Faunal Resemblance Indices (GFRIs) to several locations, finding a resemblance degree exceeding 50%. The Pikermian biome emerged approximately 8.7 million years ago, reaching its peak around 7 million years ago, and eventually disappeared around 5.3 million years ago, likely due to increasing seasonality and climatic changes linked to oceanic water circulation in the North Atlantic [16,18,55]. Pikermian species have also been reported outside the Greco-Iranian bioprovince, with occurrences from China to Spain and limited finds in Africa [51,55–58].

The extension of the Pikermian biome at Cessaniti opens intriguing considerations, strictly related to paleogeography.

4.3. Paleobiogeography

The central Mediterranean paleogeography is difficult to reconstruct for the complex geology of the area. Cessaniti is in the Capo Vaticano–Monte Poro area, between the Calabrian Arc and the Tyrrhenian Basin. The Capo Vaticano promontory has been separated from the Serre massif by the opening of the Mesima graben during the Miocene–Pleistocene and has been strongly affected by Quaternary tectonics [59,60]. The transgressive event of the late Miocene affected a land whose extension and position are not known. The Cessaniti mammal assemblage is indicative of a land having relationships with the Greco-Iranian and possibly the Afro-Arabian bioprovinces. This land had to be connected extensively enough to support large mammals with specific ecological needs. The Cessaniti land was not isolated, as the absence of endemic characteristics in mammals indicates, and had not had faunal exchanges nor territorial continuity with the two bio-provinces identified in Italy (Tusco-Sardinia and Apulo-Abruzzi) and the site of Gravitelli in Sicily:

- The fossil record from the Tusco-Sardinian bioprovince is notably endemic and impoverished, largely due to a prolonged period of insular conditions [61–64].
- The Apulo-Abruzzi area featured strongly endemic mammals since the early Tortonian [61,65–68].
- The Gravitelli mammal assemblage (Messina, Sicily) is based on fossils no longer available to the study, because they were destroyed during the earthquake that hit the city of Messina in 1908. The revision of available museum casts and the papers by Seguenza [69,70] revealed a non-endemic fauna with European affinity and possibly two African taxa [71,72]. The re-consideration of the stratigraphy described by Seguenza could indicate an age more ancient than the Messinian age originally attributed, maybe older than 7 Ma [71]. The site is no longer investigable for the intense urbanization. The mammal association does not have taxa in common with Cessaniti, except for *Ceratotherium* sp.

The paleogeographic scenarios for the Cessaniti land are as follows:

- Land connected to North Africa: in this case, the Greco-Iranian taxa come from an expansion to Africa of the Pikermian biome; this widely hypothesized condition is not strongly supported by evidence.
- Land connected to Eastern Europe: in this case, the land was subject to complex movements, maybe related to the Adria plate; this hypothesis is not supported by geological data.

5. Conclusions

The attribution of the giraffid remains here ascribed to *Samotherium boissieri*, in addition to *Bohlinia attica* and *Tragoportax* cf. *rugosifrons*, confirms the similarities of the Cessaniti faunal assemblage to the Greco-Iranian bioprovince.

The presence of *Ceratotherium* and *Stegotetrabelodon* would suggest an Afro-Arabian component in the assemblage. However, the presence of a *Stegotetrabelodon syrticus* with plesiomorphic characters leads to a revision because the specimens from Cessaniti have been compared to the whole variability of the species, which included the Arabian specimens recently attributed to *Stegotetrabelodon emiratus*.

The mammal assemblage of Cessaniti reveals an area without connection to other central Mediterranean lands. The original position and connections of the land of Cessaniti in the Mediterranean during the late Miocene is still unknown.

Funding: This research received no external funding.

Institutional Review Board Statement: Not applicable.

Informed Consent Statement: Not applicable.

Data Availability Statement: The original contributions presented in this study are included in the article. Further inquiries can be directed to the corresponding author.

Acknowledgments: The author is deeply indebted to the past superintendents Fabrizio Sudano and Michele Mazza (SABAP of Reggio Calabria; Ministry of Culture, Italy) for allowing this research; to Vincenzo Calzona, director of the MuRi, for the access to collections; to Maurizio Paoletti, president of the SiMU, and Adriano Guido, responsible for the SiMU Section of Paleontology, the University of Calabria, for the access to collections. The Geosciences staff for the kindness in every step of the process; last but not least, the anonymous reviewers for their helpful suggestions which improved the paper.

Conflicts of Interest: The author declares no conflicts of interest.

Abbreviations

The following abbreviations are used in this manuscript:

MuRi	City Museum of Ricadi, Via Roma, 12, 89866 Santa Domenica, Vibo Valentia, Italy
SiMU	Museal System of the University of Calabria, Section of Palaeontology, Via Pietro Bucci 87036 Rende, Cosenza, Italy
MuMe	"Museo della Memoria", Via Trento 2, Parghelia, Vibo Valentia, Italy
GPT	"Gruppo Paleontologico Tropeano", last address at the "Civico Museo del Mare di Tropea", vico Ospedale n. 1 89861—Tropea (VV) Vibo Valentia, Italy

References

1. Gramigna, P.; Guido, A.; Mastandrea, A.; Russo, F. The Paleontological Site of Cessaniti: A Window on a Coastal Marine Environment of Seven Million Years Ago (Southern Calabria, Italy). *Geol. Romana* **2008**, *41*, 25–34.
2. Marra, A.C.; Carone, G.; Agnini, C.; Ghinassi, M.; Oms, O.; Rook, L. Stratigraphic and Chronologic Framework of the Upper Miocene Cessaniti Succession (Vibo Valentia, Calabria, Italy). *Riv. Ital. Paleontol. Stratigr.* **2017**, *127*, 379–393.
3. Checchia-Rispoli, G. Illustrazione Dei Clipeastri Miocenici Della Calabria Seguita Da Uno Studio Sulla Morfologia Interna e Sulla Classificazione Dei Clipeastri. In *Parte 3 di Memorie per Servire alla Descrizione della Carta Geologica d'Italia*; Tipografia della Pace E. Cuggiani: Roma, Italy, 1925; Volume 9.
4. Carone, G.; Domning, D.P. *Metaxitherium serresii* (Mammalia: Sirenia): New Pre-Pliocene Record, and Implications for Mediterranean Paleoecology before and after the Messinian Salinity Crisis. *Boll. Soc. Paleontol. Ital.* **2007**, *46*, 55.
5. Guido, A.; Marra, A.C.; Mastandrea, A.; Tosti, F.; Russo, F. Micromorphological, Geochemical, and Diagenetic Characterization of Sirenian Ribs Preserved in the Late Miocene Paleontological Site of Cessaniti (Southern Calabria, Italy). *Facies* **2012**, *58*, 179–190. [CrossRef]
6. Ferretti, M.P.; Rook, L.; Torre, D. *Stegotetrabelodon* (Proboscidea, Elephantidae) from the Late Miocene of Southern Italy. *J. Vertebr. Paleontol.* **2003**, *23*, 659–666. [CrossRef]
7. Ferretti, M.P.; Rook, L.; Carone, G.; Marra, A.C. New Findings of *Stegotetrabelodon syrticus* from the Late Miocene of Cessaniti, Southern Italy. *Boll. Soc. Paleontol. Ital.* **2017**, *56*, 89–92.
8. Ferretti, M.P. Miocene Proboscideans from Italy: African Elements and Palaeogeographic Implications. *Geol. East Libya* **2008**, *3*, 325–334.
9. Pandolfi, L.; Marra, A.C.; Carone, G.; Maiorino, L.; Rook, L. A New Rhinocerotid (Mammalia, Rhinocerotidae) from the Latest Miocene of Southern Italy. *Hist. Biol.* **2021**, *33*, 194–208. [CrossRef]
10. Marra, A.C. Out of Pikermi: The Occurrence of *Bohlinia* in the Late Miocene of the Central Mediterranean. *Geosciences* **2025**, *15*, 44. [CrossRef]
11. Marra, A.C.; Solounias, N.; Carone, G.; Rook, L. Palaeogeographic Significance of the Giraffid Remains (Mammalia, Arctiodactyla) from Cessaniti (Late Miocene, Southern Italy). *Geobios* **2011**, *44*, 189–197. [CrossRef]
12. Marra, A.C. *Tragoportax* cf. *rugosifrons* (Schlosser, 1904) from the Late Miocene of Cessaniti (Southern Italy). *Comptes Rendus Palevol.* **2018**, *17*, 378–387.
13. Marra, A.C. Contribution of the Late Miocene Mammals from Calabria and Sicily to the Palaeogeography of the Central Mediterranean. *Atti Della Accad. Pelorit. Pericolanti Cl. Sci. Fis. Mat. Nat.* **2019**, *97*, 29.
14. Kostopoulos, D. The Late Miocene Mammal Faunas of the Mytilinii Basin, Samos Island, Greece: New Collection 13. Giraffidae. *Beitr. Zur Palaeontol.* **2009**, *31*, 299–343.
15. Solounias, N. *Putting Samotherium in Its Place: The Morphology of Giraffids and the Geology of Samos*; Cambridge Scholars Publishing: Newcastle upon Tyne, UK, 2024; ISBN 9781036404147.
16. Witmer, L. The Paleoecology of the Pikermian Biome and the Savanna Myth. In *Hominoid Evolution and Climatic Change in Europe: Volume 1, The Evolution of Neogene Terrestrial Ecosystems in Europe*; Solounias, N., Plavcan, J.M., Quade, J., Witmer, L., Eds.; Cambridge University Press: Cambridge, UK, 1999; Volume 1.
17. Solounias, N.; Rivals, F.; Semprebon, G.M. Dietary Interpretation and Paleoecology of Herbivores from Pikermi and Samos (Late Miocene of Greece). *Paleobiology* **2010**, *36*, 113–136. [CrossRef]
18. Kostopoulos, D.S. The Pikermian Event: Temporal and Spatial Resolution of the Turolian Large Mammal Fauna in SE Europe. *Palaeogeogr. Palaeoclimatol. Palaeoecol.* **2009**, *274*, 82–95. [CrossRef]
19. Mayor, A. *The First Fossil Hunters: Paleontology in Greek and Roman Times*; Princeton University Press: Princeton, NJ, USA, 2000; ISBN 0691089779.

20. Mayor, A. The 'Monster of Troy' Vase: The Earliest Artistic Record of a Vertbrate Fossil Discovery? *Oxf. J. Archaeol.* **2000**, *19*, 57–63. [CrossRef]
21. Monge-Nájera, J. Evaluation of the Hypothesis of the Monster of Troy Vase as the Earliest Artistic Record of a Vertebrate Fossil. *Uniciencia* **2020**, *34*, 147–151. [CrossRef]
22. Solounias, N.; Mayor, A. Ancient References to the Fossils from the Land of Pythagoras. *Earth Sci. Hist.* **2004**, *23*, 283–296. [CrossRef]
23. Nicotera, P. Rilevamento Geologico Del Versante Settentrionale Del Monte Poro (Calabria). *Mem. Note Dell'istituto Geol. Appl. Napoli* **1959**, *7*, 1–92.
24. Marra, A.C.; Somma, R.; Guido, A. The Palaeontological Heritage of the Capo Vaticano–Monte Poro Area (Vibo Valentia, Italy): Research, Protection, and Management. *Acta IMEKO* **2023**, *12*, 1–11. [CrossRef]
25. Carone, G.; Domning, D.P.; Marra, A.C. New Finds of *Metaxytherium serresii* (Gervais, 1847) (Mammalia: Sirenia) from the Upper Miocene of Monte Poro (Calabria, Italy). *Boll. Della Soc. Paleontol. Ital.* **2013**, *52*, 187–196.
26. Gramigna, P.; Bassi, D.; Russo, F. An Upper Miocene Siliciclastic-Carbonate Ramp: Depositional Architecture, Facies Distribution, and Diagenetic History (Capo Vaticano Area, Southern Italy). *Facies* **2012**, *58*, 191–215. [CrossRef]
27. Gramigna, P.; Morsilli, M.; Russo, F. Facies Changes in Siliciclastic-Carbonate Depositional Environments of the M.te Poro Area (Upper Miocene, Southern Calabria). *Rend. Online Soc. Geol. Ital.* **2008**, *2*, 93–96.
28. Caracciolo, L.; Gramigna, P.; Critelli, S.; Calzona, A.B.; Russo, F. Petrostratigraphic Analysis of a Late Miocene Mixed Siliciclastic-Carbonate Depositional System (Calabria, Southern Italy): Implications for Mediterranean Paleogeography. *Sediment. Geol.* **2013**, *284–285*, 117–132. [CrossRef]
29. Marra, A.C.; Sudano, F.; Rao, A.; Calzona, V.; Guido, A. Promotion of Paleontological Heritage: Case Histories from Southern Italy. *Heritage* **2025**, *8*, 100. [CrossRef]
30. Von Den Driesch, A. *A Guide to the Measurement of Animal Bones from Archaeological Sites*; Peabody Museum Press: Cambridge, MA, USA, 1976; Volume 1, ISBN 0873659503.
31. Bohlin, B.; Sefve, I.; Ringström, T.J.; Zdansky, O. *Die Familie Giraffidae Mit Besonderer Berücksichtigung Der Fossilen Formen Aus China*; Geological Survey of China: Beijing, China, 1926; Volume 4.
32. Kostopoulos, D.; Koufos, G.; Koliadimou, K. The Giraffids (Mammalia, Artiodactyla) from the Late Miocene Mammalian Localities of Nikiti (Macedonia, Greece; *Palaeontogr. Abt. A* **1996**, *239*, 61–88.
33. Koufos, G.D.; Kostopoulos, D.S.; Vlachou, T.D. The Late Miocene Mammal Faunas of the Mytilinii Basin, Samos Island, Greece: New Collection. 16. Biochronology. *Beitr. Zur Paläontol.* **2009**, *31*, 397–408.
34. Ríos, M.; Danowitz, M.; Solounias, N. First Comprehensive Morphological Analysis on the Metapodials of Giraffidae. *Palaeontol. Electron.* **2016**, *19.3.50A*, 1–39. [CrossRef]
35. Solounias, N.; Danowitz, M. Astragalar Morphology of Selected Giraffidae. *PLoS ONE* **2016**, *11*, e0151310. [CrossRef]
36. Roussiakis, S.; Iliopoulos, G. Preliminary Observations on the Metrical Variation of *Helladotherium duvernoyi* and *Bohlinia a Attica*. In Proceedings of the 5th International Symposium on Eastern Mediterranean Geology, Thessaloniki, Greece, 14–20 April 2004; pp. 343–346.
37. Xafis, A.; Mayda, S.; Grímsson, F.; Nagel, D.; Kaya, T.; Halaçlar, K. Fossil Giraffidae (Mammalia, Artiodactyla) from the Early Turolian of Kavakdere (Central Anatolia, Turkey). *Comptes Rendus Palevol.* **2019**, *18*, 619–642. [CrossRef]
38. Geraads, D. Remarques Sur La Systématique et La Phylogénie Des Giraffidae (Artiodactyla, Mammalia). *Geobios* **1986**, *19*, 465–477. [CrossRef]
39. Geraads, D. Les Giraffinae (Artiodactyla, Mammalia) Du Miocène Supérieur de La Région de Thessalonique (Grèce). *Bull. Muséum Natl. D'histoire Nat. Sect. C Sci. Terre Paléontol. Géol. Minéral.* **1979**, *1*, 377.
40. Geraads, D.; Spassov, N.; Kovachev, D. Giraffidae (Artiodactyla, Mammalia) from the Late Miocene of Kalimantsi and Hadjidimovo, Southwestern Bulgaria. *Geol. Balc.* **2005**, *35*, 11–18. [CrossRef]
41. Geraads, D. Les Gisements de Mammifères Du Miocène Supérieur de Kemiklitepe, Turquie: 8. Giraffidae. *Bull. Muséum Natl. D'histoire Nat. Sect. C Sci. Terre Paléontol. Géol. Minéral.* **1994**, *16*, 159.
42. Gentry, A.W.; Rössner, G.E.; Heizmann, E.P.J. Suborder Ruminantia. *Miocene Land Mamm. Eur.* **1999**, *225*, 258.
43. Bernor, R.L.; Rook, L. A Current View of Sahabi Large Mammal Biogeographic Relationships. *Garyounis Sci. Bull. Spec. Issue* **2008**, *5*, 285–292.
44. Sanders, W.J. Proboscidea from the Baynunah Formation. In *Sands of Time: Ancient Life in the Late Miocene of Abu Dhabi, United Arab Emirates*; Springer: Berlin/Heidelberg, Germany, 2022; pp. 141–177.
45. Bibi, F.; Kaya, F.; Varela, S. Paleoecology and Paleobiogeography of the Baynunah Fauna. In *Sands of Time: Ancient Life in the Late Miocene of Abu Dhabi, United Arab Emirates*; Springer: Berlin/Heidelberg, Germany, 2022; pp. 333–350.
46. Whybrow, P.J.; Hill, A.P. *Fossil Vertebrates of Arabia: With Emphasis on the Late Miocene Faunas, Geology, and Palaeoenvironments of the Emirate of Abu Dhabi, United Arab Emirates*; Yale University Press: New Haven, CT, USA, 1999.

47. Bibi, F. Bovidae and Giraffidae from the Baynunah Formation. In *Sands of Time: Ancient Life in the Late Miocene of Abu Dhabi, United Arab Emirates*; Springer: Berlin/Heidelberg, Germany, 2022; pp. 219–241.
48. Solounias, N.; McGraw, W.S.; Hayek, L.-A.; Werdelin, L. The Paleodiet of Giraffidae. In *Antelopes, Deer and Relatives*; Yale University Press: New Haven, CT, USA, 2000.
49. Ríos, M.; Sánchez, I.M.; Morales, J. A New Giraffid (Mammalia, Ruminantia, Pecora) from the Late Miocene of Spain, and the Evolution of the Sivathere-Samothere Lineage. *PLoS ONE* **2017**, *12*, e0185378. [CrossRef]
50. Spassov, N.; Geraads, D. *Tragoportax* PILGRIM, 1937 and *Miotragocerus* STROMER, 1928 (Mammalia, Bovidae) from the Turolian of Hadjidimovo, Bulgaria, and a Revision of the Late Miocene Mediterranean Boselaphini. *Geodiversitas* **2004**, *26*, 339.
51. Eronen, J.T.; Ataabadi, M.M.; Micheels, A.; Karme, A.; Bernor, R.L.; Fortelius, M. Distribution History and Climatic Controls of the Late Miocene Pikermian Chronofauna. *Proc. Natl. Acad. Sci. USA* **2009**, *106*, 11867–11871. [CrossRef]
52. Olson, E.C. The Evolution of a Permian Vertebrate Chronofauna. *Evolution* **1952**, *6*, 181–196. [CrossRef]
53. Ataabadi, M.M.; Bernor, R.L.; Kostopoulos, D.S.; Wolf, D.; Orak, Z.; Zare, G.; Nakaya, H.; Watabe, M.; Fortelius, M. Chapter 25. Recent Advances in Paleobiological Research of the Late Miocene Maragheh Fauna, Northwest Iran. In *Fossil Mammals of Asia: Neogene Biostratigraphy and Chronology*; Columbia University Press: New York, NY, USA, 2013; pp. 546–565.
54. Eronen, J.T.; Fortelius, M.; Micheels, A.; Portmann, F.T.; Puolamäki, K.; Janis, C.M. Neogene Aridification of the Northern Hemisphere. *Geology* **2012**, *40*, 823–826. [CrossRef]
55. Kurtén, B. *The Chinese Hipparion Fauna: A Quantitative Survey with Comments on the Ecology of the Machairodonts and Hyaenids and the Taxonomy of the Gazelles*; Munksgaard: Copenhagen, Denmark, 1952.
56. Casanovas-Vilar, I.; García-Paredes, I.; Alba, D.M.; van den Hoek Ostende, L.W.; Moyà-Solà, S. The European Far West: Miocene Mammal Isolation, Diversity and Turnover in the Iberian Peninsula. *J. Biogeogr.* **2010**, *37*, 1079–1093. [CrossRef]
57. Bibi, F. Mio-Pliocene Faunal Exchanges and African Biogeography: The Record of Fossil Bovids. *PLoS ONE* **2011**, *6*, e16688. [CrossRef] [PubMed]
58. Böhme, M.; Spassov, N.; Majidifard, M.R.; Gärtner, A.; Kirscher, U.; Marks, M.; Dietzel, C.; Uhlig, G.; El Atfy, H.; Begun, D.R. Neogene Hyperaridity in Arabia Drove the Directions of Mammalian Dispersal between Africa and Eurasia. *Commun. Earth Environ.* **2021**, *2*, 85. [CrossRef]
59. Cirrincione, R.; Fazio, E.; Fiannacca, P.; Ortolano, G.; Pezzino, A.; Punturo, R. The Calabria-Peloritani Orogen, a Composite Terrane in Central Mediterranean; Its Overall Architecture and Geodynamic Significance for a Pre-Alpine Scenario around the Tethyan Basin. *Period. Mineral.* **2015**, *84*, 701–749.
60. Russo, D.; Fiannacca, P.; Fazio, E.; Cirrincione, R.; Mamtani, M.A. Geology and Microstructural Evolution of Syn-to Late-Tectonic Granitoids from Capo Vaticano Promontory (Southern Calabria, Italy). *Geol. Mag.* **2025**, *162*, e8. [CrossRef]
61. Rook, L.; Gallai, G.; Torre, D. Lands and Endemic Mammals in the Late Miocene of Italy: Constrains for Paleogeographic Outlines of Tyrrhenian Area. *Palaeogeogr. Palaeoclimatol. Palaeoecol.* **2006**, *238*, 263–269. [CrossRef]
62. Rook, L. Geopalaeontological Setting, Chronology and Palaeoenvironmental Evolution of the Baccinello-Cinigiano Basin Continental Successions (Late Miocene, Italy). *Comptes Rendus Palevol* **2016**, *15*, 825–836. [CrossRef]
63. Cirilli, O.; Benvenuti, M.; Carnevale, G.; Casanovas Vilar, I.; Delfino, M.; Furio, M.; Papini, M.; Villa, A.; Rook, L. Fosso Della Fittaia. The Oldest Tusco-Sardinian Late Miocene Endemic Vertebrate Assemblage (Baccinello-Cinigiano Basin, Tuscany, Italy). *Riv. Ital. Paleontol. E Stratigr.* **2016**, *122*, 13–34.
64. Abbazzi, L.; Delfino, M.; Gallai, G.; Trebini, L.; Rook, L. New Data on the Vertebrate Assemblage of Fiume Santo (North-West Sardinia, Italy), and Overview on the Late Miocene Tusco-Sardinian Palaeobioprovince. *Palaeontology* **2008**, *51*, 425–451. [CrossRef]
65. van den Hoek Ostende, L.W.; Meijer, H.J.M.; van der Geer, A.A.E. A Bridge Too Far. Comment on "Processes of Island Colonization by Oligo-Miocene Land Mammals in the Central Mediterranean: New Data from Scontrone (Abruzzo, Central Italy) and Gargano (Apulia, Southern Italy)" by PPA Mazza and M. Rustioni [Palaeogeography, Palaeoclimatology, Palaeoecology 267 (2008) 208–215]. *Palaeogeogr. Palaeoclimatol. Palaeoecol.* **2009**, *279*, 128–130.
66. Mazza, P.P.A.; Rustioni, M. Five New Species of Hoplitomeryx from the Neogene of Abruzzo and Apulia (Central and Southern Italy) with Revision of the Genus and of Hoplitomeryx Matthei Leinders, 1983. *Zool. J. Linn. Soc.* **2011**, *163*, 1304–1333. [CrossRef]
67. Mazza, P.P.A.; Rustioni, M. Processes of Island Colonization by Oligo–Miocene Land Mammals in the Central Mediterranean: New Data from Scontrone (Abruzzo, Central Italy) and Gargano (Apulia, Southern Italy). *Palaeogeogr. Palaeoclimatol. Palaeoecol.* **2008**, *267*, 208–215. [CrossRef]
68. Freudenthal, M.; van den Hoek Ostende, L.W.; Martín-Suárez, E. When and How Did the Mikrotia Fauna Reach Gargano (Apulia, Italy)? *Geobios* **2013**, *46*, 105–109. [CrossRef]
69. Seguenza, G. Nuovi Resti Di Mammiferi Pontici Di Gravitelli Presso Messina. *Boll. Della Soc. Geogr. Ital.* **1907**, *26*, 89–122.
70. Seguenza, L. I Vertebrati Fossili Della Provincia Di Messina: Parte II. Mammiferi e Geologia Del Piano Pontico. *Boll. Della Soc. Geol. Ital.* **1902**, *21*, 115–172.

71. Iannucci, A. A Reappraisal of the Lost Suids from the Late Miocene of Gravitelli (Sicily, Italy) and Paleobiogeographical Implications. *Palaeoworld* **2024**, *33*, 469–491. [CrossRef]
72. Gallai, G.; Rook, L. *Propotamochoerus* sp. (Suidae, Mammalia) from the Late Miocene of Gravitelli (Messina, Sicily, Italy) Rediscovered. *Riv. Ital. Paleontol. Stratigr.* **2006**, *112*, 317.

MDPI AG
Grosspeteranlage 5
4052 Basel
Switzerland
Tel.: +41 61 683 77 34

Life Editorial Office
E-mail: life@mdpi.com
www.mdpi.com/journal/life

www.ingramcontent.com/pod-product-compliance
Lightning Source LLC
Chambersburg PA
CBHW061620210926
51850CB00009B/2386
9783725880034